Gary E. McCuen

IDEAS IN CONFLICT SERIES

502 Second Street
Hudson, Wisconsin 54016
Phone (715) 386-7113

Library of Congress Cataloging-in-Publication Data

Militarizing space.

(Ideas in conflict)
Bibliography: p.
1. Strategic Defense Initiative. 2. Astronautics, Military—United States. I. McCuen, Gary E. II. Series.
UG743.M555 1989   358'.1754   88-42900
ISBN 0-86596-070-4

Illustration & photo credits

Jerry Fearing 44, John Houser 143, Dan Hubig 76, Gary Huck 93, Etta Hulme 81, Craig MacIntosh 37, 50, Nuclear War Graphics Project 57, Ohman 116, Mike Peters 88, Steve Sack 12, 22, 67, 108, William Sanders 16, Jerry Scott 28, David Seavey 85, 125, Stein 100, Toles 134.

© 1989 by Gary E. McCuen Publications, Inc.
502 Second Street • Hudson, Wisconsin 54016
(715) 386-7113
International Standard Book Number 0-86596-070-4
Printed in the United States of America

CONTENTS

CHAPTER 3 SDI AND SCIENTISTS

CHAPTER 4 STAR WARS AND MORAL CHOICES

REASONING SKILL DEVELOPMENT

These activities may be used as individualized study guides for students in libraries and resource centers or as discussion catalysts in small group and classroom discussions.

IDEAS in CONFLICT ®

This series features ideas in conflict on political, social, and moral issues. It presents counterpoints, debates, opinions, commentary, and analysis for use in libraries and classrooms. Each title in the series uses one or more of the following basic elements:

Introductions that present an issue overview giving historic background and/or a description of the controversy.

Counterpoints and debates carefully chosen from publications, books, and position papers on the political right and left to help librarians and teachers respond to requests that treatment of public issues be fair and balanced.

Symposiums and forums that go beyond debates that can polarize and oversimplify. These present commentary from across the political spectrum that reflect how complex issues attract many shades of opinion.

A *global* emphasis with foreign perspectives and surveys on various moral questions and political issues that will help readers to place subject matter in a less culture-bound and ethnocentric frame of reference. In an ever-shrinking and interdependent world, understanding and cooperation are essential. Many issues are global in nature and can be effectively dealt with only by common efforts and international understanding.

Reasoning skill study guides and discussion activities provide ready-made tools for helping with critical reading and evaluation of content. The guides and activities deal with one or more of the following:

RECOGNIZING AUTHOR'S POINT OF VIEW

INTERPRETING EDITORIAL CARTOONS

VALUES IN CONFLICT

WHAT IS EDITORIAL BIAS?

WHAT IS SEX BIAS?

WHAT IS POLITICAL BIAS?

WHAT IS ETHNOCENTRIC BIAS?

WHAT IS RACE BIAS?

WHAT IS RELIGIOUS BIAS?

*From across **the political spectrum** varied sources are presented for research projects and classroom discussions. Diverse opinions in the series come from magazines, newspapers, syndicated columnists, books, political speeches, foreign nations, and position papers by corporations and nonprofit institutions.*

About the Editor

Gary E. McCuen is an editor and publisher of anthologies for public libraries and curriculum materials for schools. Over the past 18 years his publications of over 200 titles have specialized in social, moral, and political conflict. They include books, pamphlets, cassettes, tabloids, filmstrips, and simulation games, many of them designed from his curriculums during 11 years of teaching junior and senior high school social studies. At present he is the editor and publisher of the *Ideas in Conflict* series and the *Editorial Forum* series.

CHAPTER 1

STAR WARS AND NATIONAL SECURITY

STAR WARS AND NATIONAL SECURITY

A VISION OF HOPE

Ronald Reagan

The following reading is the conclusion of President Ronald Reagan's March 23, 1983, speech on defense spending and defensive technology. In this speech, President Reagan announced the plans for a space-age strategic defense initiative, more commonly referred to as "Star Wars."

Points to Consider:

1. Describe President Reagan's position on the reduction of nuclear arms.
2. How would the strategic defense initiative work?
3. Summarize the President's message.

President Ronald Reagan, March 23, 1983, speech on defense spending and defense technology.

Let me share with you a vision of the future which offers hope. It is that we embark on a program to counter the awesome Soviet missile threat with measures that are defensive.

Now, thus far tonight I've shared with you my thoughts on the problems of national security we must face together. My predecessors in the Oval Office have appeared before you on other occasions to describe the threat posed by Soviet power and have proposed steps to address that threat. But since the advent of nuclear weapons, those steps have been increasingly directed toward deterrence of aggression through the promise of retaliation.

This approach of stability through offensive threat has worked. We and our allies have succeeded in preventing nuclear war for more than three decades. In recent months, however, my advisers, including in particular the Joint Chiefs of Staff, have underscored the necessity to break out of a future that relies solely on offensive retaliation for our security.

Over the course of these discussions, I've become more and more deeply convinced that the human spirit must be capable of rising above dealing with other nations and human beings by threatening their existence. Feeling this way, I believe we must thoroughly examine every opportunity for reducing tensions and for introducing greater stability into the strategic calculus on both sides.

One of the most important contributions we can make is, of course, to lower the level of all arms, and particularly nuclear arms. We're engaged right now in several negotiations with the Soviet Union to bring about a mutual reduction of weapons. I will report to you a week from tomorrow my thoughts on that score. But let me just say, I'm totally committed to this course.

If the Soviet Union will join with us in our effort to achieve major arms reduction, we will have succeeded in stabilizing the nuclear balance. Nevertheless, it will still be necessary to rely on the specter of retaliation, on mutual threat. And that's a sad commentary on the human condition. Wouldn't it be better to save lives than to avenge them? Are we not capable of demonstrating our peaceful intentions by applying all our abilities and our ingenuity to achieving a truly lasting stability? I think we are. Indeed, we must.

A Vision of the Future

After careful consultation with my advisers, including the Joint Chiefs of Staff, I believe there is a way. Let me share with you a vision of the future which offers hope. It is that we embark on a program to counter the awesome Soviet missile threat with measures that are defensive.

10

Let us turn to the very strengths in technology that spawned our great industrial base and that have given us the quality of life we enjoy today.

What if free people could live secure in the knowledge that their security did not rest upon the threat of instant U.S. retaliation to deter a Soviet attack, that we could intercept and destroy strategic ballistic missiles before they reached our own soil or that of our allies?

I know this is a formidable, technical task, one that may not be accomplished before the end of this century. Yet, current technology has attained a level of sophistication where it's reasonable for us to begin this effort. It will take years, probably decades of effort on many fronts. There will be failures and setbacks, just as there will be successes and breakthroughs. And as we proceed, we must remain constant in preserving the nuclear deterrent and maintaining a solid capability for flexible response. But isn't it worth every investment necessary to free the world from the threat of nuclear war? We know it is.

In the meantime, we will continue to pursue real reductions in nuclear arms, negotiating from a position of strength that can be ensured only by modernizing our strategic forces. At the same time, we must take steps to reduce the risk of a conventional military conflict escalating to nuclear war by improving our non-nuclear capabilities.

America does possess—now—the technologies to attain very significant improvements in the effectiveness of our conventional, non-nuclear forces. Proceeding boldly with these new technologies, we can significantly reduce any incentive that the Soviet Union may have to threaten attack against the United States or its allies.

As we pursue our goal of defensive technologies, we recognize that our allies rely upon our strategic offensive power to deter attacks against them. Their vital interests and ours are inextricably linked. Their safety and ours are one. And no change in technology can or will alter that reality. We must and shall continue to honor our commitments.

I clearly recognize that defensive systems have limitations and raise certain problems and ambiguities. If paired with offensive systems, they can be viewed as fostering an aggressive policy, and no one wants

11

Cartoon by Steve Sack. Reprinted by permission of *Star Tribune, Newspaper of the Twin Cities.*

that. But with these considerations firmly in mind, I call upon the scientific community in our country, those who gave us nuclear weapons, to turn their great talents now to the cause of mankind and world peace, to give us the means of rendering these nuclear weapons impotent and obsolete.

Tonight, consistent with our obligations of the ABM treaty and recognizing the need for closer consultation with our allies, I'm taking an important first step. I am directing a comprehensive and intensive effort to define a long-term research and development program to begin to achieve our ultimate goal of eliminating the threat posed by strategic nuclear missiles. This could pave the way for arms control measures to eliminate the weapons themselves. We seek neither military superiority nor political advantage. Our only purpose—one all people share—is to search for ways to reduce the danger of nuclear war.

My fellow Americans, tonight we're launching an effort which holds the promise of changing the course of human history. There will be risks, and results take time. But I believe we can do it. As we cross this threshold, I ask for your prayers and your support.

Thank you, good night, and God bless you.

STAR WARS AND
NATIONAL SECURITY

A SUPERHEROIC MYTH

Robert Jewett

The following reading appeared in The Christian Century, *a weekly ecumenical publication. The article responds to President Reagan's March 23, 1983, speech on defense spending and defense technology, and provides a theological perspective to the Star Wars debate.*

Points to Consider:

1. Compare and contrast the Star Wars mentality of today with the zealous nationalism of the eighth century B.C.
2. What percentage of calls and telegrams supported the President's plans for a space-age ABM system?
3. How have popular religion and superheroic stories shaped our public consciousness?
4. Can reliance on military defense achieve peace and security? Why or why not?

Although it defies every standard of common sense, the idea that creating a new generation of superweapons will change "the course of human history" and "eliminate the threat posed by strategic nuclear weapons" is given a certain plausibility by the superheroic framework of Reagan's policies.

There is a passage in Isaiah that echoes the optimism of President Reagan's March 23 ABM speech. In that speech, Reagan connected the development of new defensive technologies with the vision of a "free people" living "secure in the knowledge that their security did not rest on the threat" of atomic retaliation. Similarly, 27 centuries ago Isaiah depicted the confidence of King Hezekiah and his advisers after the strengthening of their defenses and the signing of a mutual defense treaty with the Egyptians. "We have made a covenant with death, and with Sheol we have an agreement," Isaiah portrays them as gloating to themselves. "When the overwhelming scourge passes through it will not come to us" (Isa. 28:15). While Isaiah argued that such foolish optimism was the opposite of genuine faith, he was fully aware of its alluring power. In both Isaiah's time and in our own, the impulses of popular religion have combined with a certain type of nationalism to encourage the emergence of fatal illusions.

A Nationalistic Revival

Isaiah prophesied during the nationalistic resurgence that emerged with Hezekiah's ascension to the throne of Judah in the last decade of the eighth century B.C. After a generation under the rule of Assyria, Israel, under Hezekiah's leadership, joined a new alliance with Babylon and Egypt in 705. Disregarding Isaiah's counsel, Hezekiah stopped paying the tribute of an Assyrian vassal; he fortified Jerusalem and cut the 1,700-foot waterway to the Pool of Siloam; he strengthened his army and then settled back with the assurance that "there will be peace and security in my days" (Isa. 38:8).

This nationalistic revival was congruent with fundamental impulses in Israelite religion. That God would protect the innocent who sought freedom from brutal empires was a conviction strengthened by every telling of the exodus story. That Yahweh's king should create peace and security for the innocent was a concept celebrated in the enthronement psalms as well as in prophetic oracles. And that Assyria was the eighth century "focus of evil," to use a recent expression, could hardly be doubted. It was militaristic, totalitarian, expansionistic, and heartless in its treatment of captive peoples. Dividing the world into perfectly innocent and perfectly evil nations was as easy for the people of the eighth century as it is for those of the 20th. And what person of faith can doubt that God will take the side of the righteous against the wicked?

A Massive Delusion

That Isaiah was not swayed by the zealous nationalism of his society was due to several factors that have considerable bearing on our own time. First of all, he recognized the irresistible military prowess of Assyria. No loosely organized coalition of rebels motivated by nationalistic independence stood a chance against Assyria's technique of throwing its vast military power against one small nation at a time. No fortifications could withstand the powerful siege machines the Assyrians had devised, and no elite could hope to escape the deportation that amounted to the cultural decapitation of rebellious provinces. To disregard such realities and believe in the possibility of a magical "refuge" or "shelter" was a massive delusion.

Isaiah was also concerned about the impact that Hezekiah's militaristic adventures would have on the essential tasks of government under Yahweh. The standard set for Judah's king was to "make justice the line and righteousness the plummet" (Isa. 28:17). Maintaining social justice was his essential task, with particular reference to the "poor" and the "meek of the earth." Not only was the militarization of Judean society a diversion of such a proper use of energies but its burden would fall most heavily on the poor and defenseless.

The Covenant with Death

Furthermore, the trend toward self-serving alliances, double-dealings, and nationalistic propaganda that marked Hezekiah's administration was a denial of the forthright tradition of Yahwistic "righteousness." This is why Isaiah's oracle on the "covenant with death" contains such harsh

15

'Come on, honey, I'll satisfy your wildest fantasies!'

Cartoon by William Sanders. Reprinted with special permission of NAS, Inc.

words for those whose actions bespeak the claim "we have made lies our refuge and in falsehood we have taken shelter" (Isa. 28:15). The style of propagandistic politics that takes pride in dissembling erodes the integrity of the entire society and makes honesty in the courts and in everyday affairs all the more unlikely. Isaiah therefore viewed the religious nationalism of his time as a fundamental perversion of the essential tasks of government and a violation of the will of Yahweh.

That the "covenant with death" and the "refuge" of propaganda were hollow illusions was confirmed by the invasion of Sennacherib in 701 B.C. The "overwhelming scourge" leveled 46 Judean cities and carried off more than 200,000 survivors. Jerusalem was left as the capital of a tiny vassal state while the rest of the Judean territory was resettled with non-Israelite captives and incorporated into the Assyrian provinces. The eighth century B.C equivalent of an atomic holocaust had occurred, just as the prophetic realism of Isaiah had foreseen.

Superheroic Leaders in an Imaginary World

This chilling episode has a direct relevance to Americans who are putting their faith in a new form of the "covenant with death." A remarkable outpouring of messages followed Mr. Reagan's speech announcing the plans for a space-age ABM system. Eighty percent of the calls and telegrams were supportive, according to White House adviser Michael Deaver, making this "the most favorable response to any speech since he was elected president." A clue to the plan's positive resonance with the American mind lies in the unofficial language used by the White House staff to describe the president's recent speeches. The March 8 speech to the National Association of Evangelicals in Orlando, Florida—wherein Russia was depicted as an evil empire—is referred to as the "Darth Vader" speech. The March 23 proposal to develop the ABM, the ultimate defense system in outer space, is called the "Star Wars add-on." Popular superheroic entertainment is literally the source of the moral and strategic conceptions embodied in this modern form of the "covenant with death."

The task of superheroic leaders in such stories is to protect the innocent from aggression by the exercise of superhuman powers, restoring an imaginary world of perfect security and peace. Within such a framework, it becomes plausible to concentrate on military expenditures and to exhort scientists "to turn their great talents to the cause of mankind and world peace, to give us the means of rendering these nuclear weapons impotent and obsolete."

A Mythic Perspective

Although it defies every standard of common sense, the idea that creating a new generation of superweapons will change "the course of human history" and "eliminate the threat posed by strategic nuclear weapons" is given a certain plausibility by the superheroic framework of Reagan's policies. According to this mythic perspective, a benign paradise can be regained if enough superhuman powers are granted to a selfless president. Not only is a "covenant with death" possible, but its achievement through Hezekiah-style means is consistent with America's "spiritual awakening." According to the Orlando speech, the churches' task of opposing "sin and evil in the world" involves speaking out "against those who would place the United States in a position of military and moral inferiority." If God's New Israel rearms itself sufficiently, its spiritual and military power "must terrify and ultimately triumph over those who would enslave their fellow man." Just disregard your Isaiahs and support my military budget, Mr. Reagan implies, and I will take care of the Assyrians for you.

The connection between this Star Wars mentality and the zealous nationalism of the eighth century B.C. is anything but accidental. American civil religion has been shaped by the adoption of biblical ideas and stories. Popular narratives, from the Indian captivity tales through

17

the cowboy westerns and detective stories to the superheroic redemption myths of the past 50 years, have been influenced by the same biblical tradition. These stories have made the idea of selfless battle to restore paradise a part of our public consciousness. In their basic structure and content such modern tales are similar to those that shaped the popular mind of the eighth century. In both there is a tendency to see evil focused in the enemy and to maintain an innocent, defensive, and peace-loving image of one's own martial behavior. In both there is the promise of a happy ending, victory over evil, and escape from destruction for those who are true to the right religious values. In both there is a reliance on military solutions to complex human problems.

The Effect of Heroic Tales

This final point was recently alluded to with bewilderment by George Kennan. He describes the "almost exclusive militarization of thinking and discourse about Soviet-American relations that now commands the behavior and the utterances of statesmen and propagandists on both sides of the line." Pointing to the subjective and inaccurate nature of such thinking, he refers to the "marks of an intellectual primitivism and naivete unpardonable in a great government." That sophisticated leaders of the eighth and the 20th centuries should share these traits is a sign of the alluring power of zealous nationalism sustained by popular religion and storytelling.

The effect of heroic tales about achieving a "covenant with death" is to create massive delusions about "the overwhelming scourge." Mr. Reagan speaks of the "total elimination" of "nuclear arsenals" by means of building new and better ones. Like Hezekiah, he dreams of achieving "peace and security" by expanding his military system. Such a flat defiance of common sense is, however, consistent with the *Star Wars* myth of destroying an evil empire when "the Force" directs the pushing of the button that fires the ultimate weapon and frees innocent victims from tyranny. But contrary to such stories of righteous triumph, neither the scourge of the Assyrian armies nor modern atomic exchanges can be contained or defended against. Instead, atomic war would escalate beyond all limits. The extent of one's defenses would only make the area of devastation more complete. Moreover, as the ABM treaty recognized, in the atomic era the mere prospect of developing an invulnerable defense makes pre-emptive attacks inevitable. Isaiah's realism has redoubled meaning for us today, as we realize that illusions about scientific "covenants with death" only make the death of deaths more likely.

STAR WARS AND NATIONAL SECURITY

SDI WILL RESTORE CONFIDENCE

Robert Jastrow and James Frelk

Robert Jastrow co-authored the following reading in his capacity as President of the George C. Marshall Institute. He is also founder and past Director of the NASA Goddard Institute for Space Studies.

James Frelk co-authored this reading in his capacity as Executive Director of the George C. Marshall Institute. He is also former National Security Affairs Analyst for the U.S. House of Representatives.

The George C. Marshall Institute develops educational programs relating to fields of science with an impact on public policy.

Points to Consider:

1. How did the Soviet arsenal change after 1973?
2. What is the SS-18?
3. Why does it appear that the Soviet Union is ready to break out from the ABM Treaty?
4. Compare and contrast space-based defense with ground-based "point" defense.

Robert Jastrow and James Frelk, *How the Soviets Emasculated America's Deterrent,* George C. Marshall Institute, Washington, D.C. Reprinted with permission from *Policy Review,* Summer 1987. *Policy Review* is the flagship publication of the Heritage Foundation.

Space-based defenses, even if their effectiveness is limited, have a toxic effect on first-strike planning.

In 1956, when Khrushchev threatened to intervene in the Suez crisis with nuclear rockets, North Atlantic Treaty Organization (NATO) Commander Gruenther replied, "Moscow will be destroyed as night follows day," and Khrushchev backed away. In 1973, as Soviet troops prepared to enter the Yom Kippur War, Nixon and Kissinger faced Brezhnev down with the threat of a nuclear attack.

In 1979, when Soviet troops moved into Afghanistan, Carter held a meeting with his advisers, considered a nuclear alert, and decided to withdraw from the Olympics.

The Soviet Arsenal Began to Grow

What had changed between 1973 and 1979? What was the Carter Administration afraid of? Figures on the growth of the Soviet nuclear arsenal suggest the answer. In 1973, the Soviets had no militarily effective nuclear warheads— the accurate kind that can land within a few hundred yards of a hardened missile silo or communications center and destroy it. Accurate warheads are the key to the use of a nuclear arsenal for intimidation, because if used in sufficiently large numbers in a first strike, they can cripple the other side's nuclear forces, and prevent him from effective retaliation. The United States had more than 1,000 accurate warheads in 1973. The yield of these warheads was not very large, but the Soviet targets were not very well protected either. As a consequence, we could cripple the Soviet nuclear forces, but they could not cripple ours.

But in 1974, the Soviets started deploying accurate warheads, and in 1977—a critical year of transition—they reached parity with the United States in this important category of strategic weapons. By 1979, they had 3,450 accurate weapons capable of a first strike—more than twice as many as the U.S.

The result, according to then Secretary of Defense Harold Brown, was that by 1979 the Soviet Union could destroy 95 percent of our Minuteman intercontinental ballistic missiles (ICBMs) in their silos. They could also, Secretary Brown reported, "destroy our bombers by a barrage attack . . . so that even if the bombers got off the ground, they may not escape."

When Nixon and Brezhnev signed the Anti-ballistic Missile (ABM) Treaty in 1972, the Americans thought they had obtained a promise from the USSR that it would not menace the survivability of our retaliatory forces in this way. The essence of mutually assured destruction (MAD) and the ABM Treaty, after all, was the guaranteed ability to devastate the adversary's homeland if he attacked. In fact, the Americans felt so strongly about this point they added a "unilateral understanding" to

THE ABM TREATY

Protection against accidental launches is badly needed. However, gaining this insurance will require U.S. withdrawal from the ABM treaty. Withdrawing from the ABM treaty is a good idea, because, as matters have worked out since the treaty was signed, it now leaves America vulnerable to nuclear destruction.

The Technical Panel on Missile Defense of the George C. Marshall Institute, National Review, April 1, 1988

the ABM treaty, in which they said a prime purpose of the negotiation was to "reduce threats to the survivability of our respective retaliatory forces." Failure to reach an agreement on that point, the American negotiators said, "would constitute a basis for withdrawal from the ABM treaty."

But the ink was hardly dry on the ABM Treaty when the Soviets began to slide into their silos the first of a new generation of Soviet ICBMs, more accurate than previous Soviet ICBMs, and good enough to take out our missile silos, command and control centers and other top-priority military targets. The Soviets kept on building these accurate, first-strike weapons until, in 1979, as Secretary Brown noted, they had enough of them to place at risk a large fraction of our nuclear deterrent.

A War-winning Nuclear Capability

And the Soviet arsenal continued to grow. By 1981, the number of accurate Soviet warheads had reached nearly 5,000, leading Secretary Brown to comment, in his annual report to Congress, on "the degree of emphasis in Soviet military doctrine on a war-winning nuclear capability." In one of the great understatements of all time, Dr. Brown called this development "troublesome."

The biggest of the new Soviet missiles, and the biggest ICBM in existence, is the SS-18. The SS-18 is twice as large as an MX, weighs 200 tons, is as high as a 10-story building, can carry at least 10 warheads—each with destructive power exceeding half a million tons of TNT—and has sufficient fuel in reserve to "cross-target" the entire United States. SS-18 warheads are very accurate—better than the Mark 12A warheads which have been the mainstay of the U.S. ICBM arsenal for years. The SS-18 is certainly the most fearsome weapon of mass destruction ever devised by man.

21

Cartoon by Steve Sack. Reprinted by permission of *Star Tribune, Newspaper of the Twin Cities.*

At last report, the Soviets had 308 SS-18s in the field, carrying more than 3,000 warheads. They also has 360 SS-19s and 150 SS-17s, with warheads of comparable accuracy. The number of accurate warheads known to us in this arsenal totals nearly 6,000. The destructive power residing in the deployed SS-18s alone is greater than the destructive power of the entire U.S. missile force. . . .

Soviet ABM Breakout

Anxiety over the extent of Soviet preparations for a first strike has been intensified by the discovery that the USSR also seems to be preparing a nationwide ABM defense. This discovery confronts defense planners with the frightening prospect of a Soviet first-strike force that can diminish considerably the American capacity for retaliation, and a Soviet ABM defense that could block whatever counterattack we might manage to get off the ground afterward with our crippled forces. The banning of this nightmarish possibility was, of course—for the Americans, at least— the driving force behind the ABM Treaty.

The ABM Treaty notwithstanding, technical reports received from the Defense Intelligence Agency and the CIA indicate the U.S.S.R. has acquired nearly all the elements needed for a defense against the ballistic missiles that would constitute the main instrument of American retaliation. The Soviet Union appears to be poised for a breakout from the Treaty.

Some experts say there is no cause for U.S. concern, because the Soviet ABM defense is not very effective. It is by no means as good as the defenses the SDI is designing for the 1990s; it is certainly not good enough to stop a massive U.S. first strike—if we were to launch one.

But the American arsenal being what it is, a first strike by the United States is of vanishing low probability. American defense planners are not worrying about how well Soviet ABM defenses might block a U.S. first strike; they are focused on deterring a Soviet first strike—a possibility that the Soviet ICBM buildup brings to the forefront of their attention. For that purpose, the American planners rely on the threat of massive destruction of the Soviet Union in a retaliatory second strike. And against a U.S. retaliatory second strike, with weakened and diminished forces surviving a Soviet first strike, the Soviet missile defense could be exceedingly effective. . . .

The Need for a Space-based Defense

In the ongoing, and often abrasive, argument over the Strategic Defense Initiative, this is one of the main issues that separates SDI supporters from their opponents. Opponents of SDI, by and large, do not believe that the Soviet Union can launch a successful first strike, because they do not agree that U.S. retaliatory forces are vulnerable to a surprise attack; they feel that U.S. deterrence of a Soviet attack by the threat of retaliation will remain effective for many years to come.

In our view, their confidence is contradicted by the events of the last 10 years: the new thrusts in anti-submarine warfare; the trend toward accurate, small-yield nuclear weapons; and most important, the Soviet ICBM buildup, coupled with alarming signs of Soviet preparations for an overt breakout from the ABM Treaty by the early 1990s.

As matters stand today, powerful congressional forces are opposed to the deployment of a missile defense in the 1990s. The SDI budget has been cut to levels that postpone the achievement of test objectives by several years, and a major effort is underway in Congress to force a type of compliance with the ABM Treaty that would preclude demonstrations of the first-generation space-based defenses regarded by the Department of Defense as feasible for deployment in the 1990s.

As a result of these congressional actions, it appears that the United States will have no defense—and certainly no defense based in space—against Soviet ICBMs in the 1990s. That is unfortunate, because a space-based defense located on satellites orbiting over the Soviet Union, that can shoot down the Soviet ICBMs as they rise from their silos, would have a paralyzing effect on Soviet first-strike planning. Since the planner cannot tell beforehand which missiles and warheads will be shot down and which will get through, he cannot target key sites, such as missile silos and command posts, and be confident of their destruction. Thus, the *sine qua non* of a successful first strike—the guaranteed destruction of the adversary's retaliatory forces—is denied to him. Space-based defenses, even if their effectiveness is limited, have a toxic effect on first-strike planning.

This is not true of the "point" defense favored by some members of Congress for the protection of our MX silos and other key military sites. The Soviet planner, confronted by a point defense surrounding a small number of critically important sites, can assign five, 10, or even 20 warheads to those sites to be confident of their destruction; and yet he will have consumed only a very small part of his arsenal on those targets. But if he is confronted with a space-based defense, and feels it is essential to achieve the same level of confidence in destroying these key sites, he must multiply *his entire arsenal* by a factor of five, 10, or 20, since he does not know beforehand which particular missiles in his arsenal will be shot down. Since the present Soviet arsenal cost some $700 billion, that would mean an expenditure of trillions of dollars.

This is only true of the space-based defense, not of the ground-based "point" defense. That is one of the main reasons why former Defense Secretary Weinberger has insisted on the inclusion of a space-based layer in even the earlier defenses under consideration by the Defense Department for deployment.

The outlook for the next 10 years is not promising. The congressional politics of missile defense—and especially the opposition to early deployment of a defense from prominent Members of Congress—are such that in the early 1990s the Soviet Union is likely to have a lethal combination of a first-strike attack force and a defense against retaliation—and the United States will have neither. In these circumstances, we believe it will be clear to all that the American government cannot protect its citizens from a nuclear attack, and is no longer a nuclear superpower. The consequences, writes Robert Gates, Deputy Director of the CIA, will be "awesomely negative for stability and peace." We suggest that this development will be seen by the world as the greatest military reversal the United States has ever suffered, with catastrophic political consequences certain to follow.

STAR WARS AND NATIONAL SECURITY

SDI WILL THREATEN PEACE

Robert M. Bowman

Dr. Robert M. Bowman, a retired lieutenant colonel of the United States Air Force, wrote Star Wars: Defense or Death Star? *in his capacity as President of the Institute for Space and Security Studies. The Institute for Space and Security Studies is an independent, nonprofit, tax-exempt organization devoted to research and educational activities in science and strategy relating to space and other high-technology areas important to national security and the maintenance of peace.*

Points to Consider:

1. Explain the successes and failures in the negotiations of the ABM Treaty.
2. Describe the effect of space systems on nuclear strategy.
3. How do anti-satellite (ASAT) weapons threaten our space systems?
4. Why is the United States pursuing weapons with a first-strike capability?

Robert M. Bowman, *Star Wars: Defense or Death Star?* (Chesapeake, MD: Institute for Space and Security Studies, 1985), pp. 1-7. Reprinted with permission of the Institute for Space and Security Studies.

We seem to be intent on surpassing the Soviets in the arms race in space which will make it very difficult to return space to the status of a sanctuary for peaceful and non-threatening military support systems.

The United States is unquestionably the world leader in space technology. There is presently raging a debate as to how we can use this advantage to enhance our national security. At the center of this debate is a renewal of the whole question of ballistic missile defense—an issue that was thought to have been put to rest by the ABM Treaty[1].

Space and National Security

Most strategic thinkers accept the fact that technology and military power in themselves cannot prevent nuclear war and provide for our security. They understand that security is dependent on a rational mix of the application of technology to military power and the application of diplomacy to arms control and disarmament.

Arms-control agreements in the recent past have resulted primarily in shifting the arms race to weapons not covered by the agreements. Supporters of the Freeze[2] movement point to its universality as one of its greatest virtues. Rather than limit or ban specific weapons (as has been done in the past), it attempts to put a stop to a whole range of activities connected with a broad class of weapons. It is true that, because of the breadth of the proposal, verification of it would be fairly straightforward. But there are many other types of weapons which would not be covered. It is likely that a Freeze, as presently proposed, would foreclose the arms race in the nuclear arena, only to have it accelerate in other areas, such as space weaponry.

The primary purpose for arms control is to reduce the chance of war. (Secondary benefits, like reducing the cost of preparing for war or reducing the destructiveness of war, have been rendered less important in this nuclear age.) This reading will trace the history of space weaponry and will show that preventing an arms race in space is critical to this primary arms control objective—and to national security. Allowing the arms race in space to continue would greatly increase the danger that nuclear weapons, even those remaining after a Freeze, would be used. . . .

To understand how space weapons affect the risk of war, let us review recent developments in strategic thought.

Historical Background

Public support for the Nuclear Freeze was greatly aided by the perception of the American people that we had suffered a profound and dangerous change in national policy and military strategy.

AN OBSTACLE TO ARMS CONTROL

The Star Wars program presents an obstacle to control of antisatellite weapons and nuclear arms. Pressing ahead with this program would not only foreclose effective restrictions on antisatellite weapons but would prevent serious negotiations on limiting nuclear missiles.

Howard Ris, Minneapolis Star and Tribune, *December 4, 1984*

Though divided over Vietnam, our country was for years relatively united on strategic matters. The motto of the Strategic Air Command, "Peace is our Profession," expressed the prevailing conception of our entire military effort. The military services were rather selective in the weapons they developed and deployed, choosing those which contributed to stability and rejecting those which were destabilizing and which would hurt, rather than help, the job of keeping the peace. There were always a few who cared little for strategy and yearned for whatever weaponry technology would allow, but until recently this minority had little influence.

Central to our military philosophy has been the subordination of weaponry to strategy. Our greatest success in this regard was the conclusion of the ABM Treaty in 1972. The United States and the Soviet Union both recognized that Anti-Ballistic Missile systems were potentially destabilizing. Of course, agreement on the treaty was aided by the facts that such weapons were very expensive and technically risky and that neither side perceived the possibility of emerging from an ABM race with a decided advantage. Still, the agreement was an important validation of the principle of maintaining stability in order to prevent war.

Unfortunately, the negotiations which led to this success were simultaneously our greatest failure in the subordination of weaponry to strategy, in that we refused also to outlaw MIRVs. Multiple Independently-targetable Reentry Vehicles have led directly to our present less stable situation by making a first strike theoretically advantageous. As long as there was only one warhead on each ICBM, it would take at least one ICBM to kill an ICBM. Actually, since accuracy and reliability are not perfect, the kill probability is always considerably less than 100 percent. It is about 60 percent for the new generation of highly accurate missiles. This means that if one side launches 1,000 ICBMs against 1,000 of the enemy's they will destroy about 600 of them. If both sides started with 1,000 then the attacker would be left with none, while his opponent would be left with 400 to do with as he pleased. Under such circumstances it is unlikely that either side would be foolish

27

Cartoon by Jerry Scott. Reprinted by permission of UFS, Inc.

enough to attack the other. This is a very stable situation. With MIRVs, however, a single ICBM can send two of its many warheads to each of several enemy silos, thereby destroying a number of opposing ICBMs. The newest generation can achieve about a 5 to 1 kill ratio. Thus the one to strike first can theoretically emerge with a big advantage. This destabilizing effect of MIRVs was recognized at the time, and an agreement banning them could easily have been reached. But we were blinded by our technological superiority and refused to include MIRVs in this treaty. Instead, we proceeded to put MIRVs on our missiles.

When, a few years later, the Russians followed suit, we discovered that we were less secure than before. We have created for ourselves what we now call the Window of Vulnerability—something impossible without MIRV.

The MX was supposed to solve the vulnerability problem by being highly survivable. Survivability is a highly stabilizing feature, making it possible to "ride out" a first strike and retain a strong retaliatory force. But while we were at it, we couldn't help throwing into our new missile all the goodies that advanced technology makes possible, including a highly-accurate guidance system which gives the MX a potential first strike or "silo-busting" capability. When making the MX survivable proved too expensive and difficult a task, we were left with what we have today—a system with no more survivability than its predecessors, but with much greater accuracy. Such a weapon would seem to have little use except in a first strike and thus is provocative to the other side

28

and highly destabilizing. The MX was a misfit in our deterrent strategy. Gradually, our strategy has changed to fit our weapons.

Meanwhile, war has been avoided largely because of the stabilizing influence of existing military space systems.

The Effect of Space Systems on Nuclear Strategy

The military surveillance systems of the United States and the Soviet Union have until now contributed immeasurably to peace by denying the element of surprise to an attacker and eliminating any advantage of a first strike. By giving each side the knowledge that it could not be taken by surprise, they have reduced the pressures for a pre-emptive strike and kept tensions relatively low. Space systems provide time for analysis, confirmation, consultation, and deliberation, and have made hairtrigger responses unnecessary. They have also provided the technical means of verification which have made arms control possible.

But now we are at a juncture. Space can continue to provide even greater benefits and solutions—or it can become a massive and perhaps decisive part of the problem.

What has changed?

Our military forces have become more and more dependent on space systems—not only for surveillance and warning, but also for communications, targeting, weather analysis, terrain mapping, navigation, and other "force multiplier" functions.

Once policy and strategy had been changed to accommodate the MX, and a protracted, limited nuclear exchange scenario adopted, military strategists realized to their horror that the space systems upon which their "war-fighting" capability depended were strictly peacetime systems, designed to support a strategy of deterrence and not survivable in a conflict situation. The function for which they were designed was to give early and unequivocal warning of an enemy attack and to support the launching of a retaliatory strike. It was assumed that any attempt to destroy our satellites would constitute warning that an attack was either under way or imminent and that that, in itself, would put in motion the retaliatory machinery. The obvious inability of the U.S. to keep a full set of satellite systems operating for more than a few hours into a nuclear war didn't seem to matter.

The peacetime nature of our space assets was reinforced by the decision to compel the Air Force to design all its new satellites for launch on the Space Shuttle. Over the vehement opposition of the military, the Shuttle was crammed down the throats of program offices responsible for operational satellite systems. At the time, this was deemed necessary in order to justify the Shuttle financially. And indeed, later in the development of the Shuttle, the political and financial support of the Air Force was the only thing that saved it from cancellation. Time and again, Congress was forced to ante up more money to complete the Shuttle development because of the total dependence of the Air Force upon it—a dependence that had been thrust upon the Air Force to create just this situation. The Shuttle, of course, is so vulnerable to attack, both in orbit and on the ground, and its two coastal launching sites so vulnerable, that it is inconceivable that the U.S. could launch any new or replacement satellites once any hostilities had broken out. Two WWII submarines (or rowboats for that matter) or even two terrorists with hand grenades or mortars could totally wipe out the country's Shuttle launch capability in seconds. Similarly vulnerable is our capability to communicate with the Shuttle and to get data back from it or any other satellites. Even the new multi-billion dollar Consolidated Space Operations Center (CSOC) which the Air force is building in Colorado Springs will be vulnerable to attack or sabotage by the most meager of forces.

It is therefore ironic that at the same time as national decisions were being made which irretrievably limit our space systems to the peacetime tripwire role for which they had been designed, other decisions were being made to spend hundreds of billions for weapons whose only usefulness is in a protracted nuclear war, but which depend heavily on space systems not designed to survive the start of conflict.

One choice available when this dilemma was recognized was to abandon the MX and other weapons of protracted war and to stick with a policy of war prevention. That choice was not made. Once a system gets so far in the pipeline, it is extremely difficult to kill (witness the B-1, rising from the ashes like a Phoenix). Instead, the choice made was to attempt to up-grade the nation's space systems to give them a warfighting capability.

Increasing the survivability of satellites by hardening them against attack was given much lip service and several millions of dollars, but very little was accomplished. Providing survivable launch capability by returning to expendable launch vehicles (instead of the Shuttle) was considered for selected systems. But most of the effort went into a program to develop a U.S. antisatellite (ASAT) system to match that of the Soviets. The rationale evidently was that if they're going to threaten our satellites, then we'll threaten theirs. The fact that we're much more dependent upon our satellites for command and control of strategic forces than they are did not prevent this decision from being made. The result is that ASATs now threaten to negate the stabilizing influence of surveillance and warning satellites.

For years, our policy had been to negotiate a ban on ASATs if at all possible. In 1975 we abandoned the ASAT system that we had had operationally deployed[3] since 1963. It had been a nuclear-tipped system, far too indiscriminate in its destructive power and inconsistent with our treaty[4] obligations. We recognized the fact that we were more secure in a world without ASATs than with them—even if ours were superior to the Soviets'.

The Arms Race in Space

This truth is now being ignored. We seem to be intent on surpassing the Soviets in the arms race in space and have developed a far more sophisticated, far more capable ASAT[5] than that possessed by the Soviet Union. It was ready to begin operational testing in early 1983 and had successful system tests in January and November 1984. Its first critical test against a target in space was held up by Congressional action until March 1985. Technical problems then came to light, causing further delay. The Administration, however, is determined to complete testing of an ASAT whose deployment (or absence thereof) will be almost impossible to verify. Its testing may be an irreversible step which will make it very difficult to return space to the status of a sanctuary for peaceful and nonthreatening military support systems.

As long as there are nuclear weapons and delivery systems for them, the United States and the Soviet Union are going to need space surveillance systems to provide some measure of stability. To allow those systems to be threatened by antisatellite weapons is reckless and foolhardy.

This danger is now being compounded by our unfortunate pursuit of weapons with a first-strike capability.

Some proponents of our new war-fighting strategy have invented second-strike scenarios where silo-busting capability is required, thereby justifying the MX. Others, however, blatantly talk about situations in which the U.S., in their opinion, should strike first, destroying Soviet command posts, hardened communications centers, leadership bunkers, and ICBMs in their silos. If we also abolish the ABM treaty, install a "Star Wars" missile defense system, and embark on a huge civil defense program involving evacuation of cities, we can, according to these strategists, hope to limit U.S. casualties to as few as 20 million deaths![6]

There is one minor flaw in this "optimistic" portrayal of victory. It depends on the Russians, when faced with such a capability, continuing their present policy of requiring committee approval before a nuclear strike can be ordered— a very time-consuming procedure. Clearly, if we proceed with the MX, Trident II, and Pershing II, the Soviets, with as little as four minutes warning, will have to go to an automated launch-on-warning (LOW) procedure. This would put the survival of the United States at the mercy of the reliability of Russian computers. But even our sophisticated and technologically advanced computer warning system has given many false alarms.[7] One of the recent ones was not identified as false until after six minutes had elapsed. If the Russian system did no better, such a fault would bring about the annihilation of the United States.

Administration strategists have the answer to that. "Knock out their surveillance satellites prior to a nuclear attack and they won't have any warning!'' I wonder what makes such "strategists" think the Soviets, once blinded, will just sit there and let themselves be decapitated?

Herein lies the greatest danger. Once the U.S. has both a first strike capability and an ASAT capability, what happens if a Soviet warning satellite is struck by a meteor or suffers a catastrophic electrical failure? Might they not reasonably assume that we have just destroyed their satellite in order to prevent them from seeing the attack we are launching against them? Would they not then be likely to give the order to launch a "retaliatory" attack?

First-strike offensive weapons are dangerous to our security. The ASAT is dangerous to our security. Together, they are devastating and are very likely to destroy our security by bringing on the war neither we nor the Soviets want—and which neither we nor the Soviets can survive.

---

[1] The full name of the treaty is "Treaty Between the United States of America and the Union of Soviet Socialist Republics on the Limitation of Anti-Ballistic Missile Systems." It was signed in May 1972, ratified by the Senate in August, and entered into force in October. Two years later it was modified by a Protocol limiting each side to one site instead of two.

[2] Probably the best overall description of the Nuclear Freeze is in "A Bilateral Nuclear-Weapon Freeze," by Randall Forsberg, in *Scientific American,* Volume 247, Number 5, November 1982.

[3] See "Vought Tests Small Antisatellite System," by B. A. Smith, in *Aviation Week & Space Technology,* Vol. 115, No. 19, 9 Nov 1981, pages 24, 25. Also see status in "Space Command: Setting the Course for the Future," by Edgar Ulsamer, in *Air Force Magazine,* August 1982, pages 48-55. Also, "Launch Pylon for Antisatellite System Tested," in *Aviation Week & Space Technology,* Vol. 116, No. 3, 18 January 1982; "Proposed U.S. Anti-Satellite System Threatens Arms Control in Space," by Michael R. Gordon, in *National Journal,* 31 Dec 1983; and "Anti-satellite Weapons" by Garwin, Gottfried, and Hafner, *Scientific American,* Vol. 250, No. 6, June 1984.

[4] The world's first anti-satellite test occurred in 1959, when the U.S. tested an ASAT fired from a B-47 bomber. This "Bold Orion" test was successful, with the missile intercepting a U.S. Explorer satellite. The U.S. also developed a co-orbital interceptor (similar to the current Soviet program) in the late 1950s. This SAINT (for Satellite Interceptor) program was later abandoned. From 1963 to 1967 the U.S. tested a different ASAT system, using the Army Nike-Zeus ABM missile. Starting in 1964, we tested yet another system, using the Air Force Thor missile, in tests called "Squanto Terror." The existence of this system, also called Project 437, was disclosed by President Johnson in his campaign against

Barry Goldwater. The system was declared operational in 1964 and remained deployed until 1975. Sixteen tests of Project 437 were carried out between 1964 and 1968. The destructive potential of these nuclear-tipped ASATs was first demonstrated by the "Starfish" test in 1962, when a 1.4 Megaton nuclear explosion in space destroyed several satellites, including at least one Navy navigational satellite, one AT&T communications satellite, and one British satellite. Several Secretaries of Defense testified to Congress about the operational status of Project 437. The system was abandoned in 1975 after a typhoon destroyed the launch site of Johnston Island. Meanwhile, "Bold Orion" was followed by work on another air-launched system, "Project Spike," which in turn led in the 1970s to our current Miniature Homing Vehicle (MHV) program. The first Soviet tests were a decade behind those of the United States.

5 Nuclear explosions in space (as well as those in the atmosphere or under water) are banned by the 1963 Limited Test Ban Treaty. This treaty, formally called "Treaty Banning Nuclear Weapons Tests in the Atmosphere, in Outer Space, and Under Water," prohibits *all* nuclear explosions in those regions, whether weapons-related or not. The proponents of the Excalibur system sometimes say that it is not a nuclear weapon, but merely a directed energy weapon *powered* by a nuclear explosion in space. As far as our obligations under the Limited Test Ban Treaty are concerned, it makes little difference what you call it; it is clearly prohibited.

6 "Victory Is Possible," by Colin S. Gray & Keith Payne, in *Foreign Policy,* Summer 1980, pages 14-27.

7 Of the dozens of false alerts, the information on only a few has been made public. False warnings of Soviet attack have been caused by such things as a moonrise, a flock of geese, a faulty computer chip, and human error. When, on November 9, 1979, a simulated attack test tape was fed by mistake into the operational system, NORAD was presented with an extremely realistic alert. Tactical aircraft were launched in the U.S. and Canada, and a high-level Threat Assessment Conference was convened. It took six minutes to discover that the attack wasn't real. On June 3 and again on June 6, 1980, warnings of Soviet launches of both SLBMs and ICBMs were received at SAC Headquarters and at the National Military Command Center in the Pentagon. SAC crews were scrambled and ordered to start their engines. ICBMs were brought to a high state of readiness, and airborne command aircraft took off. The cause of these two false alarms was the failure of a single #74175 chip in a Data General computer. Our knowledge of these events is due to leaks to the press. False alerts are generally not revealed to the public.

STAR WARS AND NATIONAL SECURITY

STRATEGIC DEFENSE IS NECESSARY

Edward L. Rowny

Edward L. Rowny gave the following address in his capacity as Special Adviser to the President and the Secretary of State for Arms Control Matters. Mr. Rowny spoke before the Institute for Foreign Policy Analysis Conference on the Strategic Defense Initiative (SDI), Washington, D.C., March 14, 1988.

Points to Consider:

1. Do SDI and START enhance security and stability? Why or why not?
2. How would SDI deter a nuclear first strike?
3. Describe the Soviet strategic defense effort.
4. What does the author mean when he says that the Soviets use "semantic infiltration" to oppose SDI?

Edward L. Rowny, *SDI: Enhancing Security and Stability,* United States Department of State, Bureau of Public Affairs, Washington, D.C.

The record has shown that the Soviets take arms control seriously only when it is clear that the United States is prepared to do what is required to preserve the military balance. . . . SDI promotes Soviet seriousness at the bargaining table.

In the five years since President Reagan's speech launching the Strategic Defense Initiative (SDI), we have seen great progress in the technology of strategic defenses. Indeed, the Defense Acquisition Board has approved six key SDI technologies for demonstration and validation.

But I will leave the technology to the scientists. My remarks will focus on SDI in its strategic context—how it fits in with our defense and arms control goals. I will also address how the Soviet Union responds to these goals.

SDI: Enhancing Security and Stability

Pursuing both of the approaches I discussed with him in 1980, two of the highest priorities on President Reagan's agenda are SDI and strategic arms reduction talks (START), a treaty which would reduce strategic offensive arms. The President is deeply committed to developing effective defenses against ballistic missiles and to working toward a strategic arms reduction treaty that will cut in half existing U.S. and Soviet nuclear arsenals in a manner that contributes to stability.

The common theme uniting these goals is security and stability. Future strategic defenses offer us hope against the threat of ballistic missile attack. A good START treaty will reduce that threat, too. The overarching link between these objectives is the goal of enhancing deterrence. Both seek to reduce the risk of war.

However, the popular debate on the role of strategic defense and deterrence often centers around the notion that START and SDI are competing objectives. But they must not be viewed as competitors. In fact, the United States pursues the goals in such a fashion as to make them mutually reinforcing: fewer strategic offensive weapons simplifies the task of defending against them, while the prospect of effective strategic defenses discourages Soviet reliance on their preemptive offensive nuclear strategy. This, as I will discuss later, is not the approach now taken by the Soviets.

Again, the common theme in START and SDI is enhanced security and stability. A treaty reducing the strategic nuclear arsenals of the United States and Soviet Union can contribute to the goal of enhanced security, but only if it provides for stabilizing reductions—that is, reductions in those Soviet weapons and delivery systems with the greatest first-strike potential. To use an appropriate cliche, any strategist worth his salt will tell you that it is more important to decrease first-strike incentives than it is to decrease weapons inventories. And given the Soviet record on compliance and the high national security stakes of a START agreement, it is critical that these reductions be carried out under a verification regime that provides confidence for the United States that the Soviets are complying with the agreement.

Strategic defenses meeting the stringent criteria of the United States—military effectiveness, survivability, and cost effectiveness—will contribute to those very goals of improved security and strategic stability furthered by a good START agreement.

Inhibiting Soviet First-Strike Planning

What about SDI and deterrence? It is important to recognize that deterrence can be enhanced even with a partially effective strategic defense system. This is because of the havoc such effective strategic defenses could wreak on the potential attacker's first-strike plans. Planning a nuclear first strike is a highly complicated effort with specific military objectives. Soviet military strategy has been based on the preemptive use of nuclear weapons in the effort to destroy or neutralize Western nuclear assets and to disrupt our command and control systems. Their overall objective in a nuclear war would be to deny the United States the option of effective retaliation, thereby preserving the Soviet government and their elite.

The weapon best suited to this goal is the large, MIRVed (multiple independently-targetable reentry vehicle) ICBM (intercontinental ballistic missile), which constitutes the backbone of Soviet strategic forces. Effective American strategic defenses would severely inhibit the military utility of the Soviet planner's favorite weapon and greatly contribute to the uncertainty of the Soviet attack plan. As Soviet planners come to realize the decreasing utility of the ballistic missile in this critical strategic

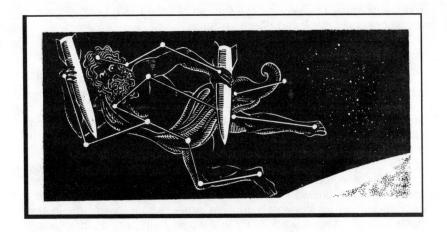

Illustration by Craig MacIntosh. Reprinted by permission of *Star Tribune, Newspaper of the Twin Cities.*

role, they will have no incentive to acquire greater numbers. Rather, they will be led to reduce their reliance on that weapon and alter their doctrine.

Denying Soviet ballistic missiles a free ride to their targets can throw a monkey wrench into the best-laid plans of the Soviet General Staff. And it is in our interest to see to it that no Soviet planner could contemplate a first strike under any circumstances with any confidence. This is what enhancing strategic deterrence is all about. SDI contributes to this goal.

Our experience since 1983 has shown that SDI has reinforced and continues to reinforce the American position at the negotiating table, especially in START. SDI played a key role in getting the Soviets back to the negotiating table in 1985 and has helped keep them there since. The record has shown that the Soviets take arms control seriously only when it is clear that the United States is prepared to do what is required to preserve the military balance. The INF (Intermediate-Range Nuclear Forces) Treaty is an excellent example of what can be achieved in arms control when the United States and its allies are ready to meet their security needs by their own action.

In the strategic arms field, continued modernization of U.S. strategic offensive forces, coupled with a vigorous strategic defense program, gets the message to the Soviets that their drive for strategic superiority will not be tolerated by the United States. SDI promotes Soviet seriousness at the bargaining table. Moreover, deployed strategic defenses would actually strengthen a START regime. While they would not decrease the importance of cheating, effective defenses could reduce its impact by providing a margin of safety as a hedge against a clandestinely deployed offensive force.

A Cooperative Transition to Defenses

Just as clearly, a good START treaty supports our goals for SDI. It's as simple as realizing that fewer offensive ballistic missile warheads— a smaller threat—make the defensive job that much easier. This is another reason we pursue a START treaty—and why we reject the Soviet effort to kill or cripple the Strategic Defense Initiative as the price of that deal.

The U.S. approach to strategic stability and enhancing deterrence is directly reflected in our arms control positions at the nuclear and space talks. We are working toward a stabilizing and verifiable 50 percent reduction in strategic offensive arms, while advancing in the defense and space talks a treaty that would help provide for predictability in the strategic relationship and for the possibility of moving cooperatively toward a more stable, increasingly defense-reliant deterrent regime.

The Soviet Union, however, has not adopted a similarly progressive approach. The Soviets would like to preserve their offensive force advantages while they pursue their own strategic defense programs. So the Soviets still maintain their linkage between START reductions and crippling restrictions on the U.S. SDI program, limits they seek to impose on SDI above and beyond those already agreed by the sides in the ABM (Anti-Ballistic Missile) Treaty. They continue to hold offensive reductions hostage to U.S. compliance with Soviet-defined limits on strategic defense work. They do this even though Soviet strategic weapons are now four times the number they were in 1972, when the United States concluded the ABM Treaty in the belief that it provided the premise for reducing the then-existing strategic offensive nuclear arsenals.

The Soviets don't impose linkage because they object to strategic defense in principle. They have their own strategic defense program, estimated to cost about $20 billion annually, the existence of which they categorically denied until General Secretary Gorbachev's offhand admission of it to Tom Brokaw.

The Soviet strategic defense effort, in fact, is comprehensive and long-standing. It consists of the permitted 100-interceptor system deployed around Moscow, which the Soviets are now upgrading; a comprehensive passive defense program for the protection of the Soviet leadership and key industry; massive strategic air defenses (over 12,000 SAM [surface-to-air missile] launchers); and programs investigating many of the same advanced strategic defense technologies under investigation in SDI.

This advanced technology program is, moreover, no "response" to SDI. Its various elements have been in place since the 1960s, and it represents, as a whole, a much greater investment of plant space, capital, and manpower than does SDI. The Soviets are investigating weapons technologies for kinetic energy, particle beam, radio-frequency, and laser weapons.

Soviet investment in their laser weapon program is especially interesting and instructive, since the Soviets so often denigrate the prospects for these advanced technology weapons. The Soviet military laser program involves some 10,000 of their top scientists and engineers and would cost us $1 billion a year to duplicate. It is centered at Sary Shagan, where the Soviets also conduct other ABM activities. The Sary Shagan facility features several air defense lasers and two lasers probably capable of damaging some components of satellites in orbit. One of these lasers is suitable for ballistic missile defense feasibility testing.

"Semantic Infiltration" Opposing SDI

It stretches one's credulity to reconcile the aggressive Soviet strategic defense program with Soviet rhetoric on SDI. The Soviets have charged that SDI is a U.S. attempt to gain strategic superiority, to generate a new round in "the arms race," to "militarize space," and to undermine the basis for offensive arms reductions. However, in Geneva, the Soviets have shown themselves unwilling to engage in open discussion of key issues, such as the nature of strategic stability, the possible contributions of defenses to stability, measures for ensuring predictability in the strategic relationship, and the offense-defense relationship.

No state is so strong a proponent of strategic defense in practice as the Soviet Union, yet none is more strongly opposed to SDI in public. Standing Soviet rhetoric side-by-side with their strategic defense efforts, one is led to conclude that the Soviets are far more interested in stigmatizing the U.S. defense effort than in engaging in a reasonable and constructive dialogue on the future of the strategic relationship and the role of strategic defenses in it.

The Soviets have recently adopted the theme that the issue in the defense and space talks is not SDI but the ABM Treaty. They have downplayed their polemical attacks on SDI in favor of arguing for "stability," which they say means an unconditional commitment to the ABM Treaty. But changes in Soviet public statements, in my judgement, reflect more of a shift in the style than in the substance of their position.

If there is one thing that Gorbachev and his new team in Moscow represent, I believe, it is the realization that one can draw more flies to honey than to vinegar. So the Soviets are practicing their time-honored technique of semantic infiltration by employing some of our lexicon to serve their political ends; for example, by emphasizing the word "stability," by which they mean the United States observing the ABM Treaty on Soviet terms, although the Soviets themselves are violating the treaty. They have toned down some of the harsher aspects of their rhetoric. But their goal remains the same—killing SDI, quickly or slowly. In their well-orchestrated public campaign of antipathy to the U.S. investigation of strategic defenses, the Soviets even trot out the very same Soviet scientists who develop Soviet strategic defense technology to allege that "it can't be done" and "it's destabilizing." This must be recognized as another cynical attempt to undermine a legitimate effort that "threatens" only a Soviet military advantage.

We must look to our own interests. The American people overwhelmingly support the idea of defense against ballistic missile attack. Yet the U.S. Congress has been unwilling to provide the funds necessary to move the program as fast as the technology will permit.

It is clear now, five years from President Reagan's visionary speech, that SDI's future depends not so much on the ingenuity of our scientists or the limits of technology as it does on our willingness to meet the political challenges posed by the possibility of effective strategic defenses. The road ahead is difficult, and we can expect the Soviets to remain uncooperative. We must demonstrate to them our resolute determination to move forward to a strategic balance incorporating defenses which threaten no one. There is no reason we should be wedded to an uneasy balance of nuclear terror. Rather, we should recognize that the incorporation of effective strategic defenses in the balance could serve to decrease both the chances and the threat of war. This is the real challenge of SDI.

STAR WARS AND NATIONAL SECURITY

STRATEGIC COOPERATION IS THE KEY

Jack Kidd

Major General Jack Kidd (USAF retired), a command pilot, led group, wing, and division formations during World War II. His career assignments have included division chief of the Organization of the Joint Chiefs of Staff, commander of the Air Force's nuclear test support unit, director of personnel planning at the Headquarters U.S. Air Force, and, upon occasion, coordinating military planning between the Joint Chiefs of Staff, State Department, and Central Intelligence Agency.

Points to Consider:

1. Describe the deficiencies of Star Wars.
2. Explain the Strategic Cooperation Initiative.
3. Summarize the partnership concept.
4. Do you believe Star Light is a possible alternative to Star Wars? Why or why not?

Jack Kidd, "'Star Light' Can Be an Alternative to 'Star Wars'," *The Humanist,* September/October 1985, pp. 27-28, 50. This article first appeared in *The Humanist* issue of September/October 1985 and is reprinted by permission.

Needed is a pervasive, long-term, coherent, consistent, bipartisan strategy for dealing with the Soviet Union.

There is still time to stop the relentless drive toward implementing the "Star Wars" program. The key is recognition that there are alternatives. The one that I am proposing—along with my associates who represent a wide spectrum of scientific and peace endeavors—is practical and developed from actual experience in military planning and study of the nuclear dilemma.

SDI Will Be a Grave Danger

The Strategic Defense Initiative (SDI), commonly referred to as "Star Wars," is a defensive concept designed to place an impenetrable shield against intercontinental ballistic missiles (ICBMs) over the United States. The logistics for the implementation of this program exist only in the minds of a few people. The hardware and software, as well as much of the basic science, does not exist. I believe that Star Wars will be unworkable. Moreover, it will be a grave danger to the United States because it suffers these deficiencies:

- Small, nuclear-tipped, low-flying cruise missiles that can underfly the defensive shield are now possessed by each side, and they can be made cheaply and in large numbers.

- Hundreds of thousands of decoys mixed with warheads can overwhelm the defenses.

- Cheap counter-measures are inevitable, leading to counter-counter-measures, ad infinitum.

- Before completion of the system, there will be a temptation for the superpower trailing to attack before being shut out. The side with the system in place will be tempted to pre-empt this attack to further reduce the threat, which would, to some degree, produce a nuclear winter. The stability of current mutual deterrence will have been lost.

- The constituency of the military-industrial complex will have become more deeply entrenched.

The devastation of England and Germany by aircraft and missile attack illustrates the low effectiveness of past defenses. Even if Star Wars were 99 percent effective in defending the United States, cities such as New York, Washington, Chicago, San Francisco, Los Angeles, and as many as 195 other targets would be destroyed, based upon current estimates of Soviet warheads, not counting future warheads. After forty years of the arms race, we are less secure today than when it began; Star Wars continues the arms race well into the next century.

Needed—A New National Strategy

Needed is a pervasive, long-term, coherent, consistent, bipartisan strategy for dealing with the Soviet Union. We propose the Strategic Cooperation Initiative (SCI), or "Star Light" strategy. This new strategy has six parts.

First, the United States should announce that it will maintain a nuclear parity with the Soviet Union and negotiate a freeze of current forces. We believe effective military superiority to be impossible in today's nuclear world.

Second, the United States should continue serious arms reduction talks to reduce numbers of ICBMs, theater missiles, and other nuclear and conventional forces and to exclude Star Wars-type systems from space.

Third, negotiate a new agreement with the Soviets on conduct of the superpowers by redefining and refining the 1972 "Basic Principles of Relations Between the U.S.A. and U.S.S.R.," which dealt with co-existence, mutual restraint, noninterference in internal affairs, and with neither taking unilateral advantage. Meanwhile, the United States must sharpen its ability to zealously discourage exporters of revolution by the Soviets or their proxies or any other nation.

Fourth, initiate with the Soviets massive joint projects to:

- Correct human environmental damage (save world ecology);

- Eliminate starvation in the world; and

- Improve global health standards.

Fifth, take part in additional projects, such as:

- Undertake joint U.S.A.-U.S.S.R. scientific ventures, including space exploration. Achieve mutual rescue capability of manned earth orbiters.

- Work with the Soviets to discover ways to alleviate hate, aggression, violence, and other human conflicts, identified recently as the greatest challenge to science. Use the new U.S. Institute for Peace to fund and coordinate this scientific endeavor.

- Continue collaboration with the Soviets on reducing nuclear proliferation.

- Improve crisis control. Facilities such as proposed by the Nuclear Negotiation Project, Harvard Law School, would assist in preventing incidents such as the downing of Korean airliner KAL 007 and would be of help in case of accidental or terrorist detonation of a nuclear weapon.

Sixth, develop a U.S. government-private sector partnership to bring into being an entire fabric of relationships with the Soviets leading to something approaching "normalcy."

- George Kennan, after fifty-six years of studying Soviet-American relations, says that this is necessary to secure arms reduction agreements or, if secured, to make them endure.

44

- Aleksandr Solzhenitsyn said that the senseless, immoral concessions by the West to the communists must stop. He adds that there is a liberation of the human spirit taking place in the Soviet Union that is the key to everything. "If we can at least slow down the process of concessions, if not stop it altogether—and make it possible for the process of liberation to continue in the communist countries— ultimately these two processes will yield us our future."

Our job is to expedite this process.

A Partnership

President Eisenhower said in 1956, "If we are going to take advantage of the assumption that all people want peace, then the problem is to get together and leap governments. . .if necessary to evade governments. . .to work out not one but thousands of methods by which people can gradually learn a bit more of each other."

A partnership would eliminate any need to leap or evade our government.

Premier Gorbachev said in May 1985, "We want to revive the spirit, the atmosphere, and the essence of detente precisely because we intend to advance even further toward a dependable system of international law and order and security."

Under the partnership concept, the U.S. government would secure renewal of all thirteen cultural, scientific, trade, and other agreements with the Soviets which have expired in the past few years. Private individuals and enterprise would be encouraged to exploit these agreements. The government would promote trade of nonstrategic goods with the Soviets and again offer Most Favored Nation treatment to the Soviets. Secretary of Commerce Baldridge was in Moscow recently proposing increased trade. My hope is that our government will encourage the Soviets to reciprocate the spirit of this partnership.

The partnership concept would encourage *private* individuals and organizations to undertake numerous actions to create a cooperative fabric of relationships between the two governments.

Across the nation there is a debate currently underway on the effectiveness of Star Wars. Star Light provides a viable alternative at this crucial time.

- Star Light goes to the root of the problem.
- Star Light can permit the arms race to wind down and arms agreements to endure.
- Star Light can lead the superpowers away from confrontation.
- Star Light can lead the way toward living together peacefully on this planet.

- Star Light can permit children to once again begin thinking about their future.

- Star Light releases the enlightened side of humans.

It was no doubt U.S. steadfastness that brought Premier Gorbachev to the point of proposing the move "toward a dependable system of international law and order and security." President Reagan should take him up on it by proposing the entire Star Light Strategy.

STAR WARS AND NATIONAL SECURITY

SDI: AN APPROPRIATE RESPONSE TO ADVERSE TRENDS

Strategic Defense Initiative Organization

The Strategic Defense Initiative Organization (SDIO) was established in response to President Reagan's March 23, 1983 speech announcing the strategic defense initiative. SDIO is coordinating research to determine whether the effort is technically feasible. This reading is excerpted from a speech given by James A. Abrahamson, director of the SDIO.

Points to Consider:

1. What kind of trends led to the pursuit of SDI?
2. Why does Mr. Abrahamson refer to the ABM Treaty as a "grand experiment"?
3. If SDI cannot provide perfect defense, what purpose will it serve?
4. How does Mr. Abrahamson suggest that we approach the Soviets with regard to SDI?

James A. Abrahamson, "Overview—The New Strategic Defense Debate," in *Strategic Defense in the 21st Century,* ed. Hans Binnendijk (Washington, D.C.: Foreign Service Institute, 1986), pp. 1-9.

The SDI grew out of a recognition of the state of technology and the long-term trends in the future.

I'm very pleased that the Foreign Service Institute has initiated this discussion on the Strategic Defense Initiative (SDI). The issue is how we can ensure that the United States, our allies, and our way of life can truly survive. How can we seek to achieve real stability, perhaps even in the face of very difficult crises in the future?

The Search for Real Stability

This new program does not imply that everything the United States has done in the past has been wrong, nor does it repudiate how we have gotten to where we are. The SDI grew out of a recognition of the state of technology and the long-term trends in the future, and that was what was behind President Reagan's March 1983 speech.

That speech caught many by surprise, especially those who considered that the strategy of mutual assured destruction, particularly in its simplest form, the form of simple retaliation, was effective for all time. But there were others who had been thinking for a long time that strategic and technological trends may not assure real survival in the next century and beyond.

Before his speech, the President had several sets of advisors. The SDI did not just come full-blown in that speech. While he did not have a careful, in-depth study by the entire administration to support his speech and to examine this change in policy, he was receiving independent advice from many technologists. The Joint Chiefs of Staff were telling him that the pendulum had swung too far in the area of offensive retaliation and that the United States should definitely begin to reconsider the strategy of defense across the entire spectrum of our security policy, not merely in the area of ballistic missiles. They urged him and have continued to urge him to reaffirm their very basic belief that the concepts and the technology that we have relied on so far are not now adequate and will not be adequate for the future.

A number of trends worry the Joint Chiefs of Staff. I'd like to mention those very briefly, and then I'd like to spin a tale about what strategic defense might look like after the year 2000.

Soviet Offensive Build-up

The first trend to consider is the offensive Soviet build-up in the context of the 1972 ABM (Anti-Ballistic Missile) Treaty. When we signed that treaty we embarked on an incredible experiment, and we embarked on it in good faith and with high hopes and expectations. At that time we as a nation agreed and the Soviets agreed, at least in form—the question is how much in substance—to leave our populations utterly defenseless to the worst and most dangerous threat that we could define at that point. Certainly a factor was that we didn't know how to defend against that threat. The experiment was that we would agree that we would remain defenseless in the hope that, by doing so, we would not threaten the other side's ability to threaten us. The hope was that we would create an atmosphere in which arms control and arms limits of all kinds could operate and people would therefore not want to build any more than just a minimum number of offensive nuclear weapons.

There were suggestions in 1972 that the ABM Treaty itself should be reconsidered if the signatories were not able to achieve certain offensive limitation agreements in a fairly short period of time—within five years. Because of that, there is an escape clause in this treaty that isn't found in many treaties. The escape clause says that the signatory nations will review the treaty every five years, so that they could withdraw from it with six months' notice if new developments that applied to their security made it prudent to do so.

This rather grand experiment has not worked. The Soviet Union and we ourselves have been in an action and reaction cycle. We have continued to build up two nuclear arsenals so that the Soviets have enough nuclear weapons to destroy all of the targets (including both civilian and military targets) in this country several times over, and we have enough to destroy not all, but many of the Soviet Union's targets, although we have a different targeting problem than the Soviet Union.

So the first trend that concerned the President and the Joint Chiefs of Staff was the continued long-term build-up of offensive weapons that was not slowed by either the ABM Treaty or by subsequent arms limitation agreements. The total megatonnage has gone up on the Soviet side. Ours has gone down in terms of deliverable megatonnage.

Soviet Defensive Efforts

The second trend is the Soviet effort toward defenses. The Soviets have maintained the world's only operational ABM system. At this point

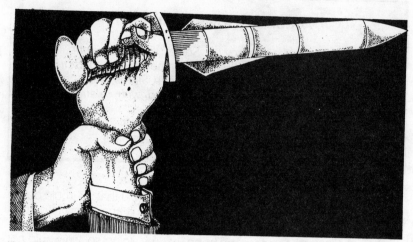

Illustration by Craig MacIntosh. Reprinted by permission of *Star Tribune, Newspaper of the Twin Cities.*

the system does not have enough weapons to defend a larger area than just a portion of the Moscow complex and the key targets that are right around that city. Nonetheless the Soviets have not just simply upgraded that system and operated it; they are building an entirely new system with huge new components. The Pushkino radar is the central command and control complex. Sketches of that particular radar show its colossal size. Two full football fields could be put on top of it. One has to wonder what its real capability is for battle management and why it was built so big. The Kransnoyarsk radar is also a serious problem. It is within treaty limits to build peripheral radars around the Soviet Union, but the Krasnoyarsk radar is not on the Soviet periphery. It's placed 4,000 kilometers in from the periphery. One has to wonder when construction on it was started. It certainly was not started after the President's 1983 speech. Planning for it probably started some time in the mid-1970s, probably shortly after the ABM Treaty was signed. The Soviets had to conduct site surveys and had to deliberately select a site inconsistent with the treaty. One has to ask why they would do that. Why would they challenge the treaty in that particular way? In my judgment it's because the location is very critical. It "closes the circle" in terms of providing a nationwide capability of radars for ballistic-missile defense. With this radar the Soviet Union could maintain accuracy in the handoff to ABM interceptor components of the more traditional kind to cover the mid-section of the country.

Thus we see that the Soviets have been working on an ABM command and control capability. We can't confirm that, of course, until those radars begin to operate. But nonetheless, battle management radars

50

were recognized in the technology of the early 1970s to be the critical element in a full ABM capability. The kinds of ABM interceptors that the Soviets have at present could easily be manufactured in mass production and could be deployed around a substantial area of the Soviet Union fairly and quickly.

Could that kind of system ever be a really effective ABM system? In my judgment, the answer is clearly no. I think that the United States could overtarget it and overcome it if we embarked on a first strike. But the Soviet ABM system could be used, if the Soviets chose to do so, very effectively against a ragged retaliatory strike. So the real threat of what the Soviets have at present is its potential for undercutting the U.S. offensive retaliatory bases for stability in today's strategy (provided the United States did not have a similar defense).

Soviet Ground-based Lasers

But the Soviets didn't stop there. They have been working for a very long time in many of exactly the same technologies that we are in the SDI. They have large ground-based lasers in place that they have been experimenting with for a very significant period, and there are many indications that they have as active a program as we do in advanced technologies.

The third trend that concerned the President and Joint Chiefs is technology itself. Right now we rely very heavily on our submarine-launched ballistic missiles. But there is no guarantee that future technology will not make this element of our strategic deterrent obsolete.

Those kinds of concerns make one ask if there is a more stabilizing kind of strategy, and the answer to that depends very heavily on what the United States can achieve technically. Can we find a way to build a very effective defense? Of course, that's what the SDI is about. It is a program that has been approved only through the research phase. People ask where research stops. It stops with the treaty, and the President has made it quite clear that the United States will limit the SDI far short of the full latitude allowed by the treaty.

The Search for a More Stabilizing Strategy

Under the treaty, we are allowed to proceed with research, just as the Soviets have, and that's what we are doing. We are looking at the kind of advanced systems that could offer an effective defense in the future. (Nothing is perfect and there is no such thing as a perfect defense.)

There is a substantial difference between the SDI and trying to defend silos. Our objective is not just to defend silos. Such an objective would continue to support the mutual assured destruction or assured retaliation doctrine. A terminal defense around those silos is only an extension of a wall around the silo that goes up into space. That is not what we are trying to do. We are trying to see if there are capabilities

51

that could allow us much more effective defenses, the kind that, while they may not be perfect, would provide stability in a crisis. These defenses would provide a deterrent capability because they would cause the Soviet targeteer to conclude that he could not assure his political leadership that a first strike would be successful from a military viewpoint. The risk associated with the lack of confidence in a first strike is part of the deterrence concept. The most important thing is that the Soviets must never be tempted to think that they can achieve any meaningful military objective with a first strike.

Our job is to build a defensive system that will create that level of lack of certainty on the part of the Soviet targeteer. The best way to do that is to build a layered defense. The United States is researching systems which might destroy a very large percentage of a strike coming from the Soviet Union or from anywhere else in the world in the boost phase, the time from launch to burnout. . . .

Laser Technology Promising

What kinds of systems is the United States working on? We have made a great deal of progress in the area of ground-based lasers. We have progressed enough in our efforts on these kinds of advanced weapons so that we are now willing to start skipping steps and possibly cut off nearly a decade from the time that was previously planned for research and development of large ground-based lasers. These lasers may become capable enough so we can fire them up through the atmosphere, bounce the beam off of a mirror, and then have that beam destroy missiles in the boost phase.

The Soviets are working on chemical lasers, carbon dioxide chemical lasers, and other older laser technologies. The laser technology that we have made the most progress in is what we call the free-electron laser, a laser in which electronics are converted to coherent light. As recently as three years ago, this was just an idea in several people's heads, and there were some very small laboratory versions. Today, we have one large free-electron laser operating out at Livermore Laboratories where we have achieved tremendous efficiencies. We have demonstrated new concepts which should enable us to step up to a very advanced and a highly accelerated development program. We will thus soon see if we are in a position to use this system to destroy missiles across great distances.

Ground-based lasers, however, are not sufficient. We need lasers that can penetrate the atmosphere. While we don't have all the answers, we recently demonstrated that we can project laser energy up through the atmosphere and correct for the distortion caused by the atmosphere. This is the result of a great deal of work that has been done over a long period of time. During the spring of 1985 we demonstrated that we could direct a high quality laser beam against airplanes at small distances. In June of 1985 we did a risk-reduction experiment against the space shuttle just to make sure our equipment worked properly. In the fall of 1985 we fired two small instrumented sounding rockets up above the atmosphere, and in fact demonstrated that at very low powers we can indeed correct for the atmosphere. The next challenge is to see if we can use these same techniques for a very high-powered laser.

Directed Energy Weapons

The next job is to see if directed energy weapons can be pointed and tracked accurately. We are about a third of the way there, but we are at the limit of what can be done from the surface of the earth. Other directed energy devices, such as particle beams, could be very effective defensive weapons and also an effective means of discriminating between decoys and warheads with nuclear material in them.* We have one of these systems operating today at the Los Alamos Laboratory. It's a small-scale model and it's certainly not cost-effective yet, but we do understand the physics and now our problem is to see if we can scale it larger and make it less expensive.

The thrust of the SDI in nearly every one of our research areas is to be able to offer to a future President, a future Congress, and a future leadership of the Western world options for defensive systems which will be cost effective and which will provide stability.

We hope to implement the SDI program in such a way that we can have a stable transition from the current offensive deterrent system to a defense-oriented system. That may be difficult. We are discussing such a transition with the Soviets and clearly they haven't bought it yet—not by any means. Currently, the Soviets are doing, I think, what is cheapest and easiest for them, and that is to help us talk ourselves out of proceeding with the technology. That's clearly the simplest way for them to maintain whatever their objectives are.

Nonetheless, I think we have to keep discussing and working with them to see if there is a way to ensure that they understand that mutual defenses will be in their interest as well as ours. That's a tall challenge. But there has been a great deal of progress in the technical side of the program. In fact, in my judgment, we can do the technical program if we resolve as a nation to do it. Our history shows that when we decide that something is truly worthwhile and we maintain the national will to proceed, technical solutions can be found. Technical solutions are the easiest side of this problem. But it has to be carefully thought through politically, strategically, and in an arms control context as well.

---

*Directed energy is focused energy in the form of particles of laser beams that can be transmitted over relatively long distances at the speed of light. A particle beam is a stream of atoms of subatomic particles.

53

STAR WARS AND NATIONAL SECURITY

SDI: THE ILLUSION OF PERFECT DEFENSE

Worldwatch Institute

The Worldwatch Institute is an independent, nonprofit research organization based in Washington, D.C. It was started in 1975 by William Dietel of the Rockefeller Brothers Fund and Lester R. Brown, the current president. Worldwatch was established to inform policymakers and the general public about the interdependence of the world economy and its environmental support systems. The Worldwatch Institute research staff analyzes issues from a global perspective and within an integrated or interdisciplinary framework.

Points to Consider:

1. Summarize the arguments used by SDI proponents and critics regarding ballistic missile defenses.
2. Would a perfect ballistic missile defense remove the potential for an enemy attack against the United States? Why or why not?
3. Describe the new goals of SDI research. How do these goals differ from President Reagan's original SDI proposal?
4. How would an American SDI deployment affect Soviet/United States relations?

Worldwatch Institute, "Assessing SDI," *State of the World 1988* (New York: W. W. Norton and Company, 1988), pp. 137-38, 140-41, 149-50. Reprinted with permission of the Worldwatch Institute.

A growing scientific chorus holds that SDI is highly unlikely to provide a near-perfect defense at any cost.

The nightmare of nuclear weapons makes dreams of perfect defenses against them understandable. An all-out nuclear exchange would almost certainly end Soviet and American societies as we know them. Even a limited nuclear war would kill an estimated 30 million Americans and Soviets, throw both economies into indefinite decline, and cause millions of cancer deaths and genetic defects. Thus, when President Reagan's Strategic Defense Initiative (SDI) seemed to offer an alternative to Mutual Assured Destruction, it received serious attention.[1]

Assessing SDI

President Reagan at first proposed a goal of "eliminating the threat posed by strategic nuclear missiles." His original SDI plan was to develop a near-perfect defense of the entire U.S. territory. But it has become increasingly clear that defending populations against nuclear attack by a determined foe is virtually impossible. Any ballistic missile defenses deployed this century would not protect people directly; they would protect weapons.[2]

Defending weapons, according to SDI proponents, would reduce the threat of nuclear war and therefore benefit everyone. SDI critics counter that less-than-perfect defenses would, at best, cost billions of dollars and benefit no one. At worst, they argue, missile defenses on one side would create a first-strike advantage, thus increasing the temptation for both sides to launch attacks in a serious crisis.[3]

A defense deployment in the absence of new arms control measures would in any case guarantee a new arms race. The Soviets would try to catch up with the United States—and vice versa—by adding defenses of their own or, more likely, enough new warheads to overwhelm any defense. Former Secretary of Defense Caspar Weinberger voiced this likelihood when he said in 1984 that "even a probable (Soviet) territorial defense would require us to increase the number of our offensive forces."[4]

Concern that a costly SDI deployment could erode the superpowers' economic security has arisen alongside strategic matters. Some critics, for example, worry that an expensive arms race would divert capital and attention from problems of declining U.S. competitiveness and Soviet economic inefficiency. Their concerns have some merit, for early deployment might cost Americans each year during the nineties as much as they currently invest in manufacturing.[5] Similarly, early deployment would burden the Soviet Union at a time when it might otherwise place highest priority on liberalizing its economy—perhaps the best opportunity for improving Soviet-American relations since the dawn of the nuclear age. Because SDI's direct costs and its opportunity costs are both very large, its advocates must persuade policymakers that it could substantially diminish the threat of war.

The Illusion of Perfect Defense

If the superpowers could make strategic defenses work perfectly, they would escape from the mutual hostage condition of nuclear deterrence. They could, in Ronald Reagan's words, "defend rather than avenge." But a growing scientific chorus holds that SDI is highly unlikely to provide a near-perfect defense at any cost. This realization arises from the strategic and technical realities of deterrence.[6] . . .

If a ballistic missile defense could work perfectly, it would still not remove the potential for a determined enemy to attack the United States with nuclear weapons. Both the United States and the Soviet Union have bombers and cruise missiles equipped to deliver nuclear explosives, with either side having sufficient force in strategic aircraft alone to destroy the other's society. If the Soviet Union felt really threatened, it could smuggle cargoes past U.S. borders—as drug dealers do every day—and hide nuclear weapons inside them. The hope of a perfect defense against nuclear weapons thus seems, at best, a fantasy. That is why SDI has already taken on new, less-than-perfect missions, with goals that derive from traditional nuclear warfighting strategies. This reality has in fact moved the director of SDI research, Lt. Gen. Dean Abrahamson, to say, "Nowhere have we stated that the goal of the SDI is to come up with a 'leakproof' defense."[7] . . .

56

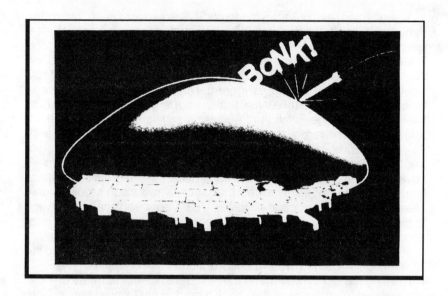

The Real World of SDI

Nuclear weapons are so powerful and diplomacy so weak that the United States and the Soviet Union stand ready to commit murder-suicide on a national—perhaps global—scale in order to avoid domination. Would that it were true that a strategic defense initiative could end for all time the terrible realities of the nuclear age.

But the Reagan vision of perfect defense is an illusion. The technology is too remote, the mission too complex, the possibilities for defeating or circumventing ballistic missile defenses too numerous. The original mission for SDI is no longer taken seriously by mainstream analysts.

The new missions that SDI has taken on are more troubling, however. They raise the spectre of new arms races and imbalances that could make crises much more dangerous. Early U.S. deployment of SDI to defend nuclear weapons would void the Anti-Ballistic Missile Treaty and push the Soviets to add their own defense, many new offensive warheads, or both. The United States and the Soviet Union would be obligated to spend hundreds of billions of dollars each. The result would, at best, make Americans and Soviets poorer.

At worst, an early deployment of SDI would create a far more dangerous world. It could make Soviet submarines vulnerable for the first time. The United States could even achieve the ability to intercept Soviet land-based missiles surviving a U.S. first strike. The possibility that the Soviet Union would be unable to retaliate after a U.S. attack would make the Soviets more trigger-happy, more likely to find advantage in a nuclear first strike. Now that the superpower relationship has thawed a bit, the negotiation of nuclear arsenals may be both useful and possible. The removal of medium-range missiles from Europe and Asia is a modest step forward. The 50-percent reduction in strategic weapons discussed at the Reykjavik summit could also improve superpower relations, ease tensions, and reduce risks of war somewhat, though no one should imagine that such a reduction would make Mutual Assured Destruction obsolete. Until wiser heads can find a way to circumvent the need for nuclear weapons altogether, balance and accommodation between the nuclear powers is essential. As Princeton foreign policy analyst Daniel Deudney has recently pointed out, too little attention is being given to such options.[8]

The singular act of reaffirming the Anti-Ballistic Missile Treaty of 1972 would avoid the arms race that an SDI deployment would cause. The treaty was drafted and signed to avoid exactly these kinds of instabilities and risks. It states in plain English that testing of new devices outside the laboratory is prohibited.[9] This constraint was the goal—indeed the language—of the U.S. negotiators. The Soviets did object to the restriction, but, more important, they signed the treaty and have obeyed it in that regard.

The Soviets have indeed violated the treaty by building a phased-array radar at Krasnoyarsk, well within their territorial boundary. The treaty prohibited such radars, which could be used to direct an SDI-like system, except on the borders where they would be less effective for use in ballistic missile defenses. Still, in an exchange of thousands of warheads when perhaps only one would destroy the radar (and therefore a ballistic missile defense system dependent on it), the radar poses little threat to U.S. security.

The deployment of SDI technologies would affect the other nuclear powers, but how is not clear. If a defense worked well enough that one superpower could not destroy the arsenals of the other, but imperfectly enough that cities could be attacked, then France, India, Israel, Pakistan, and the United Kingdom would be more equal with the United States and the Soviet Union in nuclear might. That is, their nuclear threats would be more on a strategic par with those of the superpowers.

More important, perhaps, a new arms race caused by SDI-like deployments could threaten the renewal of the Nuclear Non-Proliferation Treaty, signed in 1968 to restrict the spread of nuclear weapons. Many nations were persuaded to forgo arms development in part in return for a commitment by the superpowers to reduce their own arsenals. The treaty expires in 1992; renewal will be difficult enough because the United States and the Soviet Union have vastly expanded their arsenals. If arms expansion were to accelerate, renegotiation of this important treaty could be problematic.

At bottom, the best hope for avoiding nuclear war lies in changing fundamentally the connection between the United States and the Soviet Union. The relationship might actually mature if General Secretary Gorbachev succeeds in liberalizing the Soviet Union, and if American leaders become astute enough to recognize and capitalize on the opportunity. Deploying an SDI system could extinguish this hope.

---

[1] U.S. Congress, Office of Technology Assessment (OTA), *The Effects of Nuclear War* (Washington, D.C.: U.S. Government Printing Office, 1979); Samuel Glasstone and Philip J. Dolan, *The Effects of Nuclear Weapons* (Washington, D.C.: U.S. Government Printing Office, 1977).

[2] Reagan quote from "Weekly Compilation of Presidential Documents," Monday, March 28, 1983.

[3] For a summary of the pro and con arguments, see American Physical Society (APS), *Science and Technology of Directed Energy Weapons* (New York: 1987).

[4] Caspar Weinberger, Memorandum for the President on "Responding to Soviet Violations Polity (RSVP) Study," quoted in *Washington Post,* November 18, 1985.

[5] A space-based chemical rocket interceptor system capable of destroying submarine-launched missiles in boost-phase would cost two thirds of $1 trillion, or $66 billion per year over 10 years. By comparison, the United States recently has invested about $62 billion per year in manufacturing; see U.S. Department of Commerce, *Statistical Abstracts of the United States, 1987* (Washington, D.C.: U.S. Government Printing Office, 1987).

[6] APS, *Science and Technology of Directed Energy Weapons.*

[7] Written reply to question related to Abrahamson's testimony before the House Appropriations Committee, May 9, 1984.

8 See Daniel Deudney, "Realism's Eclipse of Geopolitics and the Loss of Strategic Bearings" (draft), Princeton University, Princeton, N.J., mimeographed, June 1987. See also Hilary F. French, "Of Nations and Nukes: The Failure of International Atomic Energy Control, 1944-1946," Honors Thesis, Dartmouth College, Hanover, N.H., May 26, 1986.

9 Article V section 1 reads "Each party undertakes not to develop, test, or deploy ABM systems or components which are sea-based, air-based, space-based, or mobile land-based"; *Treaty Between the United States of America and the Union of Soviet Socialist Republics on the Limitation of Anti-Ballistic Missile Systems,* Moscow, May 26, 1972. The definition of antiballistic missiles systems was agreed to include systems already existing or under development and "systems based on other physical principles. . .created in the future"; see *Agreed Statements,* appended to treaty.

WHAT IS POLITICAL BIAS?

This activity may be used as an individualized study guide for students in libraries and resource centers or as a discussion catalyst in small group and classroom discussions.

Many readers are unaware that written material usually expresses an opinion or bias. The skill to read with insight and understanding requires the ability to detect different kinds of bias. Political bias, race bias, sex bias, ethnocentric bias, and religious bias are five basic kinds of opinions expressed in editorials and literature that attempts to persuade. This activity will focus on political bias, defined in the glossary below.

Five Kinds of Editorial Opinion or Bias

SEX BIAS—The expression of dislike for and/or feeling of superiority over the opposite sex or a particular sexual minority

RACE BIAS—The expression of dislike for and/or feeling of superiority over a racial group

ETHNOCENTRIC BIAS—The expression of a belief that one's own group, race, religion, culture, or nation is superior. Ethnocentric persons judge others by their own standards and values

POLITICAL BIAS—The expression of political opinions and attitudes about domestic or foreign affairs

RELIGIOUS BIAS—The expression of a religious belief or attitude

---

Guidelines

Read through the following statements and decide which ones represent political opinions or bias. Evaluate each statement by using the method indicated.

Place the letter [P] in front of any sentence that reflects political opinion or bias.

Place the letter [N] in front of any sentence that does not reflect political opinion or bias.

Place the letter [S] in front of any sentence that you are not sure about.

_____ 1. Star Wars will not eliminate the threat of strategic nuclear weapons.

_____ 2. Strategic cooperation, not strategic defense, is needed to deal with the Soviets.

_____ 3. SDI will protect the American people from a Soviet first strike.

_____ 4. SDI is a politically motivated project and has little to do with science.

_____ 5. SDI promotes Soviet seriousness at the bargaining table.

_____ 6. America seems to be intent on surpassing the Soviets in the arms race in space.

_____ 7. SDI is an appropriate response to Soviet defensive efforts.

_____ 8. Space weapons are morally wrong.

_____ 9. President Reagan proposed the Strategic Defense Initiative as an alternative to mutually assured destruction.

_____ 10. SDI is a legitimate defensive weapons program.

_____ 11. Space-based defenses, even if their effectiveness is limited, have a toxic effect on first-strike planning.

_____ 12. "Star Wars" is a popular name for the Strategic Defense Initiative.

Other Activities

1. Locate three examples of political opinion or bias in the readings from Chapter One.

2. Make up one statement that would be an example of each of the following: *sex bias, race bias, ethnocentric bias, and religious bias.*

3. See if you can locate any factual statements in the twelve items listed above.

CHAPTER 2

THE COST OF SDI: MILITARY VS. ECONOMIC SECURITY

9 THE COST OF SDI

THE ECONOMIC BENEFITS OF SDI

Klaus P. Heiss

Dr. Klaus P. Heiss wrote the following article for Conservative Digest *in his capacity as president of ECON Incorporated. In his article, Dr. Heiss maintains that the benefits of an immediate deployment of the Phase I system for the Strategic Defense Initiative would be cost-effective.*

Points to Consider:

1. Does the ABM Treaty restrict deployment of Phase I of SDI? Why or why not?
2. How much will Phase I deployment cost?
3. Describe the four benefits of Phase I system deployment.
4. Compare and contrast the reasons why the Soviet Union, the United States, Europe, and Japan would benefit from deployment of Phase I systems.

Klaus P. Heiss, "The Cost to Deploy Phase One Defenses," *Conservative Digest,* May 1987, pp. 87-94. Reprinted with permission of *Conservative Digest.*

The benefits of an immediate Phase I deployment make it both cost-effective and mandatory.

The cost of a space-deployed Phase I system for the Strategic Defense Initiative (SDI) with a limited 50 percent effectiveness is estimated at $30 billion, and for a Phase I system of 90 percent (or better) effectiveness at about $100 billion. The means are known and make use of dramatic advances in kinetic-energy technologies first proposed by Lt. General Daniel O. Graham and his team at High Frontier in the early 1980s.

Among the benefits of such Phase I deployment are: 1) savings from the foregone development of such new offensive systems as Midgetman and MX railtracks ($30 billion to $100 billion); 2) possible drastic reductions in the replacement of offensive nuclear-missile forces if Phase I deployment is coupled with an arms-reduction agreement with the Soviet Union ($300 billion to $600 billion); and, 3) anywhere from $10 billion to $1 trillion in savings from the intercept of a single random nuclear missile fired by "third parties" in the 1990s.

These benefits make the immediate deployment of Phase I of SDI both cost-effective and mandatory.

In addition, Phase I deployment of SDI would compel the Soviets to scrap their deployed intercontinental ballistic missile (ICBM) force—a $1 trillion investment— and agree to nuclear disarmament or be forced to incur the enormous further costs of a hardened, fast-burn missile deployment of uncertain effectiveness.

And, finally, the proposed Phase I deployment of kinetic technology systems can proceed entirely within "the original intent" and letter of the Antiballistic Missile (ABM) Treaty.

Technology and Treaty

In the early 1980s the organization known as High Frontier developed a strategic-defense concept against nuclear missiles based primarily on relatively crude, known, and cost-effective kinetic energy technologies deployed on the ground and in space. This concept led directly to President Reagan's proposal on March 23, 1983, of a Strategic Defense Initiative.

Since that time, the technical progress on defense technologies against ballistic missiles based "on other principles" as defined in Agreed Statement D of the 1972 ABM Treaty has been so drastic that the United States can proceed today with a Phase I deployment of SDI that is based on that original "primitive" kinetic technology system proposed in the early 1980s.

Such deployment can take place within the original ABM Treaty constraints which "allowed" deployment at two U.S. sites. The plain English (or Russian) language of Article II clearly defines what "ABM" and its three components "currently" (1972) are: Component 1—missiles to shoot down ballistic missiles; Component 2—launch tubes for such missiles; and, Component 3—radars for such missiles. None of this is required under Phase I of SDI.

The consistent Soviet interpretation of the meaning of "ABM" throughout the 1970s leaves no doubt that "ABM" did not mean slingshots, bows and arrows, buckshots, lasers, particle beams, machine guns, etc. Hence such systems are not covered by Article V, which restricts deployment of "ABM" on land, the open seas, and space. Just to make sure, Article V specifically lists "missiles" and "launchers" as the "ABM" weapons to be restricted—to the *exclusion* of technologies based on "other physical principles."

With no legal obstacles in the way of immediate deployment of Phase I of SDI based on physical principles other than ABM interceptor missiles, the issue of SDI comes down to the simple question of whether such deployment is a good economic bargain for the United States. . . .The question is not whether SDI works technically—it does— but whether it is economic to deploy SDI. Here are some arguments why President Reagan's Strategic Defense Initiative and Phase I deployment make good economic sense.

Cost of Phase I Deployment: $30 Billion-$100 Billion

Depending on the level of effective protection for the United States and its allies against random nuclear-missile threats of the 1990s and limited or all-out Soviet attacks, a kinetic-technology SDI system can be deployed now within the letter of the ABM Treaty, using interceptor missiles from the two sites (originally) specified in the treaty, and SDI components based on other physical principles can be deployed elsewhere on land, the open seas, and in space per the Agreed Statement D discussions.

66

Cartoon by Steve Sack. Reprinted by permission of *Star Tribune, Newspaper of the Twin Cities.*

Such a system will cost about $30 billion if a 99 percent or better protection is desired for random (accidental, suicidal) and limited nuclear-missile attacks, and 50 percent effectiveness against all-out Soviet nuclear-missile attack. It consists of a terminal defense making use of appropriate interceptor missile technology allowed under the ABM Treaty (such as ERIS and HEDI), and ground-, sea-, and space-based kinetic technology SDI systems (such as cloud guns, space-based "garage" satellites, etc.).

This system can be expanded to provide 90 percent or better protection against all-out Soviet missile attack with a substantially increased space capability for about $100 billion.

Phase I of SDI can later be upgraded to include more-advanced technology components, such as lasers, particle beams, etc., should these technologies mature and prove to be cost-effective. . . .

While the technology and the costs of Phase I deployment of SDI are fairly well known, what are the benefits of Phase I deployment to justify a $30 billion investment of the taxpayers' money?

The limited deployment of Phase I of SDI may bring about 90 percent or more of all the benefits of SDI in the best case. Indeed, it affords the most rational approach to the problems at hand. Abstracting from the current debate on SDI (within the United States, with our allies, and with the Soviet Union) the United States and the Soviets could agree to:

- The immediate deployment of Phase I ($30 billion) by the United States, coupled with an understanding that the Soviet Union would upgrade its already-deployed SDI system around Moscow to a limited nationwide coverage. The U.S. expenditures would at best match the expenditures of the Soviets on their deployed SDI system (done under the umbrella of the ABM Treaty) and the Soviets could avail themselves of the known advances in "other physical principles" components;

- Agreement to eliminate all intermediate and intercontinental ballistic missiles, with a residual "core" force in both countries of less than 10 percent of current levels allowed under SALT II;

- Agreement of other nations to join in the reduction and elimination of ballistic nuclear missiles;

- Enforcement of this nuclear-missiles reduction agreement, and protect against violations, with Phase I deployment of SDI, to be upgraded over time to deal with any limited threat after the transition/deployment period.

In such a context the benefits of Phase I deployment are massive. Also, there is no way other than such deployment to achieve the certainty of these benefits, since any "futures" contract worth the paper it is written on is only as good as the enforcement capability that goes with it: paper is no defense against violations and surprises by the Soviets or anyone else.

Benefit 1: Elimination of Midgetman and MX Railroads Saves $30 Billion to $100 Billion

Deploying 500 Midgetmen will cost a minimum of $50 billion. This cost can easily double with more massive deployment and the many "upgrades" to be expected.

Deployment of the MX Railroad program (fifty missiles), a remnant of the Carter era, would cost another $15 billion at least. The minimum necessary MX deployment in this mode is at least 100 MX and $30 billion plus in cost.

For the price of the MX Railroad, the U.S. can now deploy the limited Phase I option of SDI, and for the price of Midgetman the U.S. can deploy a 90 percent or better global SDI Phase I system against all-out nuclear attack. . . .

Benefit 2: Reduction in Current Offensive Nuclear Missile Forces Saves $300 Billion to $600 Billion

Midgetman and MX Railroads aside, Phase I deployment coupled with massive reductions of all offensive nuclear-missile forces on both sides, to say a "core" level of 10 percent of SALT II levels, will involve additional drastic savings: over 1,000 U.S. landbased nuclear missiles and 700 submarine-launched missiles are scheduled for replacement/deployment through the 1990s at a cost of $300 billion and $700 billion respectively, or a total of $840 billion. Such deployment within the framework of past "disarmament" successes of the U.S. and the Soviets is entirely warranted without the physical ability to provide for surprise treaty violations. Indeed, these force levels have been determined precisely for their presumed "deterrent" effect. . . .

The reductions in the cost of strategic missile-force deployments of $600 billion would come about gradually, and doubtless with considerable opposition from the affected military services. Hence the cost reductions might be substantially delayed or reduced, say to half of the potential reductions ($300 billion). Either way, thanks to a Phase I deployment of SDI, these reductions would contribute substantially to reducing the federal budget deficit while assuring our national security.

Benefit 3: Protection Against Random Nuclear Missile Threats in the 1990s, at $100 Billion-Plus per Threat

Neither the ABM Treaty nor SALT I nor SALT II provide any protection whatsoever against the most realistic, and ominous, nuclear-missile threat—random threats from accident, suicidal commanders, and nuclear mini-states in the 1990s. The United States, Europe, Japan, and the Soviets will require physical defenses against such threats, or be faced with untenable blackmail situations in the next decades.

Neither the U.S. nor the Soviet Union can allow themselves to be hostage to such threats from madmen or third countries. The fatal weakness of the mutually assured destruction (MAD) doctrine underlying past strategic thinking is that MAD provides no deterrence against third-country and random mad attacks or mere threats of attacks. Consider:

A study just released by the Carnegie Endowment for Peace—an institution not known to advance arguments for SDI deployment—gives a vivid description of but a few frightening events in the past: the threatened takeover of nuclear devices by the French rebels in the Algerian conflict of the 1960s (the device was destroyed); the threatened capture of nuclear devices by the Red Guards in China in the late 1960s (again the device was destroyed before harm could be done); the unilateral bombing of Iraq's nuclear installations by Israel in 1981; and, the approach by the Soviet Union to the United States for tacit approval of a Soviet preemptive strike against Red Chinese nuclear installations in 1969.

In the 1990s a dozen or so countries will have primitive nuclear-missile capabilities, and what were isolated incidents in decades past could be daily reading in the 1990s.

The Soviet Situation

Threats from third-country forces are a reality for the Soviets today. . . .The British "threat" can be effectively blunted by the Soviets with their currently deployed SDI system around Moscow, containing some 300 interceptors. But to deal effectively with the French and British "threat" together, the Soviets need about 800 such interceptor missiles. This assumes only 60 percent effectiveness. Critics of SDI have granted in testimony to Congress that current kinetic technology would achieve "only" 75 percent effective intercept.

Seeing this, opponents of SDI are left with arguments bordering on irrationality, such as: "Negotiating an arms-control treaty with the U.S.S.R. might be the best way to prevent deployment of a Soviet ABM system that undermines the European deterrent in this century as well as in the next."[1] The fact is that the Soviets already have deployed an ABM system and certainly the Soviet Union has no intention of leaving itself open to nuclear-missile threats in the 1990s and beyond from, say, "Islamic Liberation" movements.

---

[1] John Prados, Joel S. Wit, and Michael J. Zagurek, Jr., "The Strategic Nuclear Forces of Britain and France," *Scientific American,* August 1986, pp. 33-41.

70

The U.S. Situation

If any proof were needed, the Carnegie report cited above makes clear that the non-proliferation policy has failed and that in the 1990s there will be at least a dozen "nuclear capable" states, all able to combine their primitive warheads with primitive missile capabilities. To "hit" a target as large as the U.S.A. "randomly," not much guidance technology is needed: a supply of hydrocarbons, some plumbing knowhow, and natural-gas liquefaction technology will suffice to deliver such warheads with ballistic missiles over the United States. With some software of the type found in personal computers, such countries might even succeed in coming reasonably close to major U.S. metropolitan areas. . . .

Benefit 4: The Average Cost of a Random Nuclear Missile Hit

The average cost of a random nuclear-missile hit with a primitive 500KT warhead over the continental United States is estimated to be about $100 billion—with the upper limit at about $1 trillion for major metropolitan areas, and the lower limit at about $10 billion for the relatively less-populated Western states. . . .These cost estimates include damage to real estate, industrial production, financial and information services lost, and medical costs. The damage estimates do not include any value for lives lost, or income losses from injury, or pain and suffering. . . .

The European and Japanese Situation

The arguments for deployment of SDI Phase I systems apply with equal (if not greater) force to Europe and Japan. One remembers that Libya launched six Scud missiles against Italy in 1986, and you could not find one Italian newspaper in the weeks that followed which did not take the government to task for (1) not knowing beforehand about the missile launches; and (2) for being totally defenseless against such threats.

The list of likely client states for random nuclear-missile threats contains many more countries close to European and Japanese shores than to the United States. Of course an SDI Phase I system deployed by Europe and Japan would provide those areas with effective nuclear-missile defenses against limited attacks from potential strategic adversaries. And why should the technologically advanced European and Japanese economies not afford equal (if not better) defenses against such threats as the Soviets have deployed already? . . .

In Summary

The SDI and Phase I deployment of SDI enable the United States and its allies to achieve President Reagan's vision of a gradual and (eventually) total elimination of nuclear missiles, ending the threat of instant global nuclear annihilation and of constant nuclear blackmail. Phase I of SDI is a relatively small price to pay ($30 billion) when this step is accompanied by joint or unilateral ICBM reductions. The critics of SDI have thus far lacked the imagination to see how the objectives of some of the SDI opponents (nuclear-missile disarmament, for instance) can be brought infinitely closer to fruition with Phase I deployment of SDI. . . .

All of which misses the basic rationale for Phase I systems in the United States, the Soviet Union, Europe, and Japan: namely, to have a means to stop random nuclear-missile threats from third parties in the 1990s.

THE COST OF SDI

THE ECONOMIC DEFICITS OF SDI

Rosy Nimroody

Rosy Nimroody is the national security project director for the Council on Economic Priorities, a New York City-based independent nonprofit research organization. She is the author of the recently published study Star Wars: The Economic Fallout *(Ballinger/Harper & Row).*

Points to Consider:

1. Why does the SDI program lack commercial potential?
2. Compare and contrast American and Japanese high-tech research. What will happen if SDI continues to be the backbone of American efforts?
3. How will an SDI system produce "brain drain"?
4. What percentage of Americans supported a system designed to protect U.S. missiles and key military bases?

Rosy Nimroody, "The S.D.I. Drain," *The Nation,* January 16, 1988, p. 41. Reprinted with permission of The Nation Company © 1988.

Efforts to revitalize our competitiveness by reducing deficits will be choked off if the nation proceeds to buy SDI on the installment plan.

The Reagan Administration's attempt to put America's economic house in order by reining in the galloping deficit will be in vain if plans for the Strategic Defense Initiative (SDI) continue. At an estimated cost of $400 billion to $1 trillion, SDI, or Star Wars, is not only the largest and fastest-growing program in our history but, by any measure, an actual industrial policy. The Administration has already spent $9.5 billion in the past four years and proposes to spend another $37 billion in the next five years for SDI research and development alone. This country can ill afford such a diversion of scarce financial resources into closed areas of research that lack commercial potential and do little to improve industrial productivity.

Today, the government is spending over 73 percent of its national research and development funds on projects for the military—up from 50 percent in 1980. Proceeding with SDI will increase Pentagon control of scientific research. In 1986, SDI already took up a staggering 84 percent of the growth in Defense Department funds. And according to Robert Reich, an industrial policy advocate, the SDI Organization will command roughly 20 percent of high-technology venture capital over the next four years. Research in key areas ranging from very-high-speed integrated circuits to advanced computers to optics will be sponsored by the Pentagon's SDI Organization.

Lacking Commercial Potential

The schedule-driven and weapons-oriented nature of the SDI research program, however, does not bode well for advances in civilian products. With over half the annual SDI budget allocated for such experiments as the "Excalibur" X-ray laser and kinetic energy weapons tests, the program is more concerned with displaying measurable results to maintain public and Congressional support than with finding civilian uses.

Nor are the extreme performance requirements of technology in space likely to find cost-effective applications on earth. It is difficult to imagine private uses for such SDI technologies as high-energy lasers that burn holes in metal at a distance of 1,800 miles or computer chips that are hardened against nuclear radiation blasts.

Undermining American High-Tech Efforts

In contrast, Japan's funding of civilian research on artificial intelligence is designed to improve business and consumer productivity; its laser work, to corner the market for compact-disc players and fiber optic communications. If SDI continues to be the backbone of American efforts in such high-tech areas, the result could be, as a physicist at Stanford University predicts, "the demise of the U.S. commercial laser industry in the next decade."

Where innovations in such areas as artificial intelligence, laser optics, and computer software do find commercial usefulness, their exploitation will depend on the ability of companies to wrest the technology from the government's top secret files. The Pentagon's control of the dissemination of scientific findings in the name of national security has already begun to have a chilling effect on high-tech research. For example, Pentagon restrictions on disclosure of information about very-high-speed integrated circuit chips is so tight that close-up photographs may not be published.

SDI Will Produce "Brain Drain"

Finally, a decision to produce and deploy an SDI system is sure to trigger a "brain drain" that will severely damage civilian innovation in the nation's beleaguered manufacturing companies. Up to 180,000 scientists and engineers could be on the SDI payroll by the year 2005. Industries competing for this talent will face additional costs of retraining and paying salary hikes to woo them back.

The program has spawned a powerful lobby of beneficiaries in the nation's arms industry. Between fiscal 1983 and fiscal 1986, more than 74 percent of the $7.3 billion in SDI research contracts went to just twenty contractors. Five firms alone—Lockheed, General Motors, Boeing, T.R.W., and E.G.&G.—received a total of $2.4 billion. With military funding accounting for 5 percent (G.M.) to 53 percent (Lockheed) of their

75

Illustration by Dan Hubig. Reprinted with permission of Pacific News Service.

total sales, the firms' dependency on SDI for future profitability is considerable.

In a potential conflict of interest, if SDI is deployed these contractors will determine the feasibility of the program and decide when production and deployment should begin. If Pentagon plans for early deployment of a partial defense of missile silos, as revealed by a recent report from Senator William Proxmire's office, go forward, contractors may reap the benefits of multibillion-dollar orders as early as 1994. The danger is that such programmatic and economic momentum will give SDI a life of its own regardless of its technical feasibility or strategic merit.

Can We Put the SDI Genie Back into the Bottle?

Nonetheless, a Democratic Administration in 1989 might put the SDI genie back into the bottle in the same way that Jimmy Carter canceled the B-1 bomber. With the need to balance the budget and increased reluctance to abandon the ABM treaty, Congress has shown signs of resistance to going full speed ahead with SDI. Only 15 percent of Americans sampled in an April 1987 public opinion poll conducted by Cambridge Reports supported a system designed to protect only U.S. missiles and key military bases.

Under these conditions, it makes sense to reduce SDI funding to no more than $2 billion a year—the amount that the Reagan Administration had projected for military research before embarking on the program. That level of spending would comply with the ABM treaty, hedge against a potential Soviet technological breakthrough and help reduce our federal budget deficit by $27 billion in the next five years alone. And over a twenty-year period, it would save each American household up to $12,000 in taxes.

Efforts to revitalize our competitiveness by reducing deficits will be choked off if the nation proceeds to buy SDI on the installment plan. If America's economy is to make a comeback, we can no longer delay correcting the Star Wars vision.

THE COST OF SDI

SDI WILL COST LESS THAN CRITICS SAY

Ralph de Toledano

Ralph de Toledano wrote the following article for The Independent American, *an ultra-conservative newspaper that is published every two months. In his article, Mr. de Toledano maintains that a near-term strategic defense system would cost less than critics tell us and would benefit the economy.*

Points to Consider:

1. According to the Heritage Foundation, how much will a strategic defense system cost?
2. Describe the three-layered defense system.
3. How would SDI benefit the nation's economy?
4. Why are specific SDI proposals hard to come by?

Ralph de Toledano, "SDI Will Cost Far Less Than Critics Tell Us," *The Independent American,* November/December 1987, p. 4. Reprinted with permission of Copley News Service.

By spending a total of about $120 billion in 1987 dollars, the United States could build a near-term strategic defense system that would be more than 90 percent effective against a massive Soviet nuclear strike.

By spending less than 3/1,000ths of 1 percent (.0028 percent) of gross national product over the next 10 years, or a total of about $120 billion in 1987 dollars, the United States could build a near-term strategic defense system that would be more than 90 percent effective against a massive Soviet nuclear strike. So the Heritage Foundation* tells us.

And the Heritage Foundation is correct. The cost of a Strategic Defense Initiative (SDI) system will be far less than what its critics tell us. Opponents of SDI have screamed loudly that the system would cost trillions, with such "experts" as Sen. William Proxmire, D-Wis., making the loudest and phoniest noises about costs. The real experts— and Congress knows precisely who they are but doesn't like to talk to them—have set costs for a near-term SDI at a tiny fraction of what the opponents claim.

SDI Benefits

There is another factor to the cost equation. The spin-off from SDI production—that is, the benefits to the economy and technology—would run to about 15 times the money appropriated, so that in real terms, SDI would cost nothing and bring in vast real profits to the country. This again is something the "experts" on Capitol Hill don't want us to know.

Any near-term SDI system, as Heritage analyst Grant Loebs notes, would rely heavily on both existing technology and new concepts. In time it would give us both a new physics and new mathematics. What the long-term costs would be is something else, though the spin-off would easily compensate.

Three-Layered Defense

Envisioned in the system suggested by Loebs is a three-layered defense.

"The first layer," Heritage outlines, "would use space-based kinetic-kill vehicles to destroy Soviet intercontinental ballistic missiles in their

---

*Editor's note: The Heritage Foundation is a conservative research center in Washington, D.C.

'boost phase'; the second layer would use a ground-based component to destroy missiles in the mid-course of their path; and the final layer would have an estimated 3,000 interceptors for destroying missiles at the 'terminal' state."

SDI Would Lift Federal Budget Burden

In estimating costs, however, it is important to assess the savings in offensive nuclear weapons. As a near-term SDI system fell into place, the cost of maintaining our present nuclear arsenal would sharply decline. And as long-term systems were developed, the cost of offensive nuclear weapons would be reduced to a fraction of what it is today.

From the standpoint of military economy, therefore, SDI would take a tremendous burden from the federal budget— something the critics of SDI don't like to think about.

Need Specific SDI Proposal

The problem, however, lies not so much with Congress but with the Reagan administration and the Pentagon. The White House talks rapturously but vaguely about SDI and how it would protect the American people—all true. But specific proposals with spelled-out costs are very hard to come by.

For this, the Pentagon is partially responsible. Our military leaders are sharply divided over SDI—but the division has little to do with capabilities. Inter-service rivalries have come into play. And many of the military men see themselves relegated to unimportant posts if near-term and long-term SDI systems come into being. The generals—or at least some of them—are more interested in that extra ribbon than in the defense of the country—and so they leak criticism to Congress and the press.

What is needed now is "a clearly defined SDI project," Heritage analysts argue, with equally clear cost projections. Those projections

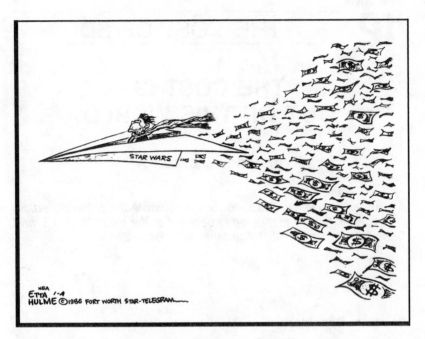

Cartoon by Etta Hulme. Reprinted by permission of NEA, Inc.

are available, but in the politically idiotic battle over SDI, no one seems to ask.

12   THE COST OF SDI

THE COST IS
OUT OF THIS WORLD

Alexander Konovalov

Alexander Konovalov, Doctor of Science, wrote the following article in his capacity as a technology specialist at the Institute of U.S. and Canadian Studies, USSR Academy of Sciences.

Points to Consider:

1. Why will SDI deepen the United States' financial problems?
2. What percentage of U.S. patents and military-related research and development find commercial uses?
3. Why have 3,500 scientists signed a petition refusing to participate in the SDI program?
4. How would U.S. allies be affected by the SDI program?

Alexander Konovalov, "Star Wars Is Economic Dead Weight," *People's Daily World,* April 4, 1987, p. 4-A.

By relying on SDI for its national security the U.S. may sacrifice its national economy.

The advocates of President Reagan's Strategic Defense Initiative (SDI), commonly known as Star Wars, claim that it will have a favorable effect on the U.S. economy, boost technological progress, open up new horizons in technology and industry, and create thousands of jobs.

These arguments have undoubtedly helped generate backing for this military program. But they are not supported by the facts.

The U.S. federal debt is now over $2 trillion. The SDI project has so far required relatively little funding, but further work on the program, estimated to cost more than $1 trillion, may begin to deepen the nation's financial problems in the very near future.

SDI Will Deepen U.S. Financial Problems

As more funds go into military programs, U.S. investment opportunities are shrinking. The net investment rate was 4 to 5 percent of the gross national product (GNP) in the 1950s and 60s, 3 to 4 percent in the 70s, and 2.8 percent in 1984. By contrast, the net investment rate in Japan is 9.8 percent.

Military-oriented projects such as SDI require both massive funding and the employment of engineers well-versed in high-tech areas. These specialists are already in short supply.

Just as any other major program, SDI will create some new jobs, but its overall impact on employment can only be negative. Studies on how military spending affects employment in U.S. industry, published in 1982, showed that every $1 billion of the taxpayers' money spent on arms resulted in a loss of 18,000 jobs.

SDI Will Not Promote Technological Progress

The argument that SDI will promote technological progress is also not supported by the facts. Major Pentagon contractors are not really interested in revolutionary scientific discoveries, which negatively affect their existing industrial capability.

More often than not, discoveries crucial to technological progress are and have been made without the Pentagon's participation. For example, the development of the Bell Laboratories' transistor, the Texas Instruments' micro-chip and the planar process, which allowed Fairchild Industries to start their mass production, were not funded by the government.

Only about half of all U.S. patents and as little as 7 percent of military-related research and development find commercial uses.

**WHAT CAN $1 TRILLION BUY
IF NOT SPENT ON STAR WARS?**

*Dr. William Sloan Coffin, Jr., recently helped put this number
into perspective:*

For one trillion dollars you could—
        build a $75,000 house
        place it on $5,000 worth of land
        furnish it with $10,000 worth of furniture
        put a $10,000 car in the garage—

*and give all this to each and every family in Kansas, Missouri,
Nebraska, Oklahoma, Colorado, and Iowa.*

*Having done this, you would still have enough left to build a
$10 million hospital and a $10 million library for each of 250 cities
and towns throughout the six-state region.*

*After having done all that, you would still have enough left out
of the trillion to put aside, at 10 percent annual interest, a sum
of money that would pay a salary of $25,000 per year for an army
of 10,000 nurses, the same salary for an army of 10,000 teachers,
and an annual cash allowance of $5,000 for each and every family
throughout that six state region—not just for one year, but forever.*

Excerpted from the Hunger Action Coalition Newsletter

Just 3 percent of the 1986 SDI budget is devoted to basic research
and many U.S. specialists fear that Army orders may undermine their
freedom to choose their own lines of research.

This is a more serious problem because spending on research with
military applications in the past six years has increased from 50 to 72
percent of the overall research and development expenditure. Commercial returns have declined accordingly.

SDI Will Not Benefit the American People

Spinoffs of military research can indeed be used for peaceful purposes, but it is hardly necessary to develop and build the B-1 bomber
just to make tennis rackets of new composite materials.

Civil aviation does not need a supersonic aircraft with a variable
geometry wing capable of flying, say, from New York to San Francisco
at tree level, remaining invisible to radar.

Direct civilian research produces far greater economic gain, as the
examples of Japan and the Federal Republic of Germany prove.

By David Seavey, USA TODAY

Cartoon by David Seavey. Copyright 1988, *USA Today*. Reprinted with permission.

The Stockholm International Peace Research Institute points out that military technology costs 20 times as much as comparable civilian systems.

Today the views of SDI proponents are widely publicized because they have powerful government backing. But this does not make their arguments more convincing, at least not for experts in the field.

Some 3,500 scientists, more than half of those doing research at 100 U.S. universities, signed a petition refusing to participate in SDI because they think it is totally unfeasible.

An official from the American Electronics Association warned that by relying on SDI for its national security the U.S. may sacrifice its national economy. The fact is that SDI may likely also damage the economies of U.S. allies involved in the project, who will also bear part of the expense.

INTERPRETING EDITORIAL CARTOONS

This activity may be used as an individualized study guide for students in libraries and resource centers or as a discussion catalyst in small group and classroom discussions.

Although cartoons are usually humorous, the main intent of most political cartoonists is not to entertain. Cartoons express serious social comment about important issues. Using graphics and visual arts, the cartoonist expresses opinions and attitudes. By employing an entertaining and often light-hearted visual format, cartoonists may have as much or more impact on national and world issues as editorial and syndicated columnists.

Points to Consider:

1. Examine the cartoon in this activity. (See next page.)

2. How would you describe the message of this cartoon? Try to describe the message in one to three sentences.

3. Do you agree with the message expressed in this cartoon? Why or why not?

4. Does the cartoon support the author's point of view in any of the readings in this book? If the answer is yes, be specific about which reading or readings and why.

5. Are any of the readings in Chapter Two in basic agreement with this cartoon?

THE ONLY THINGS THE STAR WARS DEFENSE SYSTEM WILL DEFINITELY KNOCK OUT..

...JOB CORPS, STUDENT AID, MEDICARE, ETC...

Cartoon by Mike Peters. Reprinted by permission of UFS, Inc.

CHAPTER 3

SDI AND SCIENTISTS

13 SDI AND SCIENTISTS

EARLY DEPLOYMENT: MORE POLITICS THAN SCIENCE

Union of Concerned Scientists

The following reading appeared in Nucleus, *the quarterly newsletter of the Union of Concerned Scientists (UCS). The UCS is an organization of over 100,000 scientists and citizens that coordinates a nationwide network of scientists to support arms control. The UCS established the Action Network in 1981 to involve scientists throughout the United States in the sustained and coordinated program to reduce the threat of nuclear war.*

Points to Consider:

1. Why do SDI advocates support early deployment?
2. What are SBKKVs and what function do they serve?
3. How would the Soviet Union counter the SBKKV defense?
4. Describe ERIS and HEDI.

Michael Brower, "Early Deployment: More Politics Than Science," *Nucleus,* Summer 1987, pp. 1, 4. Reprinted with permission from the Union of Concerned Scientists.

The purpose of the early deployment plan is to force the abolition of the ABM Treaty and to commit future administrations to the SDI program regardless of technical merit.

In its brief lifetime, the Strategic Defense Initiative (SDI), the administration's program to develop space-based defenses against Soviet intercontinental ballistic missiles (ICBMs), has grown into the largest item in the Pentagon's budget. Yet the program still lacks a clear purpose. It has been variously characterized as sheltering the American people from nuclear attack, protecting U.S. silo-based missiles so that they could retaliate after an attack, and providing insurance against Soviet cheating on some future arms control agreement.

As confusion and doubt about the program have grown, SDI advocates have increasingly called for early testing and deployment of a defense system based on current technologies such as rocket interceptors and radars. No technical breakthroughs have occurred since the Defense Department rejected a similar proposal as unpromising in the early 1980s. Rather, the purpose of the early deployment plan is to force the abolition of the ABM Treaty and to commit future administrations to the SDI program regardless of technical merit.

Boost-Phase Defense

In theory, a ballistic missile defense would consist of several layers, some based in space, that would intercept missiles during each phase of their flight. During a missile's boost phase, all the warheads it carried could be destroyed at one blow. But after the last booster stage stopped burning, each booster would multiply into many targets— perhaps 10 warheads and hundreds of decoys. During midcourse, in the vacuum of space, the warheads and decoys would follow the same path, and discriminating between them would be difficult or impossible until they reentered the earth's atmosphere. Thus the key to the success of a defense would be the boost-phase layer. Without an effective boost-phase layer, the mid-course and terminal layers would be overwhelmed in a large Soviet attack.

Several types of laser and particle beams have been considered for boost-phase defense. But as a result of political pressures for early deployment, the program's research focus has now shifted from these directed energy weapons to the less promising but more mature kinetic energy weapons—that is, rocket interceptors that destroy by impact. The heart of the early deployment plan is the Space-Based Kinetic Kill Vehicle, or SBKKV, for boost-phase defense. Ground-based rockets, to intercept missiles during late mid-course and reentry, are also being considered for early deployment.

SBKKVs would be deployed on "garages," or platforms, orbiting at an altitude of about 500 kilometers. Each 500-pound rocket would carry a non-nuclear homing vehicle guided by an infrared sensor, which would seek out and collide with a Soviet booster. The first generation of such rockets could be ready for testing by the late 1980s or early 1990s. (Testing SBKKVs against actual boosters, however, would violate the ABM Treaty.) SDI director General Abrahamson has suggested initially deploying 3,000 SBKKVs on 300 platforms at an estimated cost of $40-60 billion. A recent report by the conservative Marshall Institute advocates deploying 11,000 SBKKVs on about 2,000 platforms by the mid-1990s, with a cost optimistically estimated at $68 billion.

Examining SBKKV Effectiveness

How effective would a fleet of SBKKVs be against current and projected Soviet ICBM forces? At any one time, most of the orbiting SBKKVs would be out of range of Soviet launch sites. As Soviet missiles are currently deployed, the ratio of SBKKVs out of range to those within range (called the "absentee ratio") would be about 24:1, according to a study by the Congressional Office of Technology Assessment. Thus, for example, with 3,000 SBKKVs in orbit, only 125 would be within range of Soviet missile fields at any one time.

92

THESE ARE THE VOYAGES OF THE STRATEGIC DEFENSE INITIATIVE "STAR WARS." IT'S MISSION: TO EXPLORE NEW WORLDS, TO BOLDLY GO WHERE NO ONE HAS EVER BEEN BEFORE...

Cartoon by Gary Huck. Reprinted with permission of Huck/Konopacki, UE News.

If the probability that an SBKKV would successfully intercept a Soviet booster were, say, 85 percent, then 125 SBKKVs could intercept 106 Soviet boosters, carrying an average of 4.6 warheads each. Only 7.5 percent of the current Soviet force of 6,420 ICBM warheads would be eliminated. Even this estimate is optimistic, however, since it credits SBKKVs with a reliability comparable to that of ICBMs. It does not take into account the difficulties of tracking and colliding with a booster hidden by its large exhaust plume, nor such potential Soviet countermeasures as varying the acceleration of the boosters or changing the booster plume's characteristics.

Moreover, the Soviet Union is already replacing some of its older missiles with mobile, single-warhead SS-25s, and plans to deploy 10-warhead SS-24s, both in silos and on rail cars. The Central Intelligence Agency (CIA) projects that by 1994 the Soviet strategic force will comprise 1,408 missiles carrying 10,942 warheads. Against this modernized Soviet force, the proposed SBKKV fleet would be even less effective.

Even though more warheads, on average, would be eliminated with each booster destroyed, a higher absentee ratio would reduce the overall effectiveness of the SBKKV force. New Soviet boosters burn out more quickly and at lower altitudes. As a result, an SBKKV must be closer to the launch site if it is to reach the booster before it burns out; this in turn means that more SBKKVs would be out of range at any given time. In other words, the administration's projected SBKKV defense is already losing ground against projected upgrading of Soviet offenses— even if the Soviets undertake no counter-measures.

The Soviets Would Use Countermeasures

But of course the Soviet Union would take steps against U.S. defense that threatened to erode the effectiveness of its strategic arsenal. The simplest measure would be to accelerate the offensive buildup. To counter the SBKKV defense, the Soviet Union would have to add only enough new missiles and warheads to make up for those that might be destroyed, and it would have a lead time of several years in which to do so. The CIA predicts that without arms control restrictions the Soviet Union could increase the number of warheads on its missiles and bombers to 16,000-21,000 by the mid-1990s—easily enough to make up for those intercepted by the Marshall Institute defense system. Further, as the Soviet Union continued to shift to mobile missiles, it could concentrate its forces in a small area of the country, so that nearly all SBKKVs would be out of range at any one time.

Fast-burn boosters would be a second option. Present U.S. and Soviet ICBM boosters burn out in three to five minutes. Boosters could be built, however, which burn out in as little as 50 seconds, while still within the upper atmosphere. SBKKVs, with a maximum speed of about 5 kilometers per second, simply could not reach fast-burn boosters in time to intercept them. Soviet fast-burn technology has already been demonstrated in the Gazelle ABM interceptor. A recent study by the American Physical Society (APS) shows that, contrary to claims by SDI officials, such boosters could carry nearly as many warheads as current missiles.

Finally, the Soviet Union could develop a variety of methods in the next 10 years to attack SBKKV platforms in orbit. Ground-launched interceptors like those SDI is planning for late midcourse defense would be cost-effective antisatellite weapons. Only a fraction of the SBKKV platforms in orbit would have to be destroyed to clear the skies over the Soviet Union for a long enough period to launch a nuclear strike.

Still cheaper would be space mines, which would hover close to SBKKV platforms until ordered to explode. These relatively simple devices would probably be easier to develop and more reliable than ground-launched interceptors, and they could attack all platforms simultaneously. They could also be fused to explode when tampered with.

Since, as the APS study concludes, the United States is some 10 years away from knowing whether directed energy technologies will be feasible or practical for ballistic missile defense, and 20 to 30 years away from possible deployment, no replacement for SBKKV defenses would be available until long after the SBKKVs had been rendered ineffective by Soviet countermeasures.

ERIS and HEDI

Of the ground-launched rockets being considered for early deployment, the Exoatmospheric Reentry-vehicle Intercept System, or ERIS, is the focus of weapons research and development for midcourse defense. ERIS is a long-range rocket that would intercept warheads shortly before they reentered the atmosphere. An early version of this system was tested in the Homing Overlay Experiment in 1984, and an ERIS prototype could be ready for testing by about 1991. Some early deployment proposals call for several thousand ERIS interceptors. The infrared sensor on ERIS would be unable to discriminate between warheads and even the simplest decoys. Thus ERIS could easily be overwhelmed in a large Soviet attack.

The High Endoatmospheric Defense Interceptor, or HEDI, is somewhat behind ERIS in development because of the problem of atmospheric friction heating the homing vehicle's window and blinding the infrared sensor. HEDI could in principle intercept incoming warheads at an altitude of up to about 80 kilometers, depending on the height at which the ground-based radar controlling HEDI could discriminate between warheads and decoys.

During reentry, atmospheric drag would at some point cause the flight path of the lighter decoys to deviate from the path of the warheads. Nevertheless, the terminal defense might not be able to discriminate between warheads and decoys at a high enough altitude to prevent the destruction of cities. Advanced decoys now under development by the United States are designed to imitate the path and radar image of warheads to a relatively low altitude. Maneuvering reentry vehicles, which may be tested by the United States by the end of the decade, could also evade the HEDI rockets.

95

Conclusion

The current push for early deployment of SDI is no indication that it has become either technically more promising or strategically more desirable. The SBKKV in particular is an ambitious and costly weapon system, of very limited effectiveness, which would be obsolete soon after it was deployed. What the SBKKV's rising star does demonstrate is the administration's determination to leave SDI firmly entrenched when it leaves town.

SDI AND SCIENTISTS

EARLY DEPLOYMENT:
MORE SCIENCE THAN POLITICS

Edward Teller

Dr. Edward Teller is a retired nuclear physicist. He played a central role in the Manhattan Project of World War II and subsequently in the development of the hydrogen bomb. He is interviewed in this reading by John Rees, a senior editor for Conservative Digest.

Points to Consider:

1. Why does Dr. Teller support early deployment of SDI?
2. Describe Dr. Teller's position on testing of proposed anti-missile defenses.
3. Does Dr. Teller believe the Soviets are ahead of the U.S. with respect to exotic technologies? Why or why not?
4. Summarize Dr. Teller's argument to advance SDI.

John Rees, "Dr. Edward Teller Says It Is Time to Begin Deployment," *Conservative Digest,* May 1987, pp. 51-56. Reprinted with permission of *Conservative Digest.*

To start to deploy any system makes it more real. It moves then from the realm of ideas and research into concrete existence.

Years beyond what many consider retirement age, Dr. Edward Teller, now age 79, actively uses his incomparable experience and profound learning to enrich our understanding and knowledge of issues that involve a volatile mixture of science and politics. Early in his long and outstanding career as a nuclear physicist, Dr. Teller played a central role in the Manhattan Project of World War II and subsequently in the development of the hydrogen bomb. In public life, he has repeatedly served our Presidents in advisory posts on atomic energy and national defense. He maintains a pace of speaking engagements, congressional testimony, meetings, conferences, and research that would exhaust men half his age. And Dr. Teller has been a leader in galvanizing support among the Administration, the Congress, and the public for application of new developments in theoretical and applied physics for the construction of a national strategic defense that would deter the threat of nuclear first strike by the Soviet Union.

Q. Dr. Teller, as you know, there is an intense debate under way concerning near-term deployment of the Strategic Defense Initiative. Do you think it would be effective?

A. To start to deploy any system makes it more real. It moves then from the realm of ideas and research into concrete existence. However, in the public discussions so far, I do not think I have yet heard those in authority speak in concrete terms. You ask if "it" will be effective, and I must respond that I do not yet know precisely what "it" is.

But let me tell you this: I am happy that deployment is being considered because before we deploy something we really have no idea what difficulties we are going to encounter or exactly how much it will cost. If you plan to put up a factory, you first have a pilot project. Before we can have a big SDI deployment, we have to deploy parts of it. The Soviets have built parts of their SDI and deployed an anti-ballistic missile system around Moscow. To that extent, they are ahead of us. We debate, yet we have deployed nothing.

Q. Isn't there a political advantage to getting elements of SDI up now so that later Administrations can't just roll over on this issue? Is that a valid consideration?

A. Look, as I have said, to start to deploy makes things more real. It also becomes a fact, the existence of which is relevant for the next Administration. But, quite apart from any of that, there are some things which should be done soon. Not all of this can be done at once, and we must have priorities.

Q. What would head your list of priorities?

A. I can talk about a few possibilities. Many of our allies are very much interested in defense against short-range weapons. This is so not only in Europe but, quite obviously, also in Israel which is threatened by SS-21 missiles of Soviet manufacture which are installed in Syria. If Israel had a defense, we believe, those rockets could be made ineffective and the Israelis could live in peace. If not, the consequences in the short term might prove quite telling.

Does that defense against Soviet short-range missiles in Syria have anything to do with defending the United States? The answer is yes.

What the Israelis are planning to use against short-range rockets, we can use against rockets launched from fifty miles or 100 miles off our coasts from submarines. It can be used against rockets launched almost anywhere from the ocean. Today, such rockets launched from submarines or even surface ships could be very dangerous to us. That

Cartoon by Stein. Reprinted by permission of NEA, Inc.

is an important part of what we have to defend ourselves against. We have to start our defense with something.

Q. Do you think we should wait until we have an SDI defense that is 100 percent effective?

A. I don't really believe any defense will ever be 100 percent effective. But, today, we have no defense at all. And this will be a good place to start. There are many possibilities. Some people argue that we should begin to deploy appropriate kinetic-energy weapons in space. We might do that. It could be the beginning of something bigger, something more effective. And if it is, we better know how much that something bigger will cost and have opened the paths for a move into space. Pre-deployed weapons are useful only for defense; to have them in space for *offensive* purposes is obviously a waste of money. But if you know that some of them will be within a few hundred miles of a rocket launch, then we can use them to attack and stop rockets while they are accelerating and very vulnerable.

Q. Critics say that there is no way of testing in advance to see if any of the proposed anti-missile defenses would work. Is that so?

A. It is our responsibility to get experience with such systems to assist our work in defense. The Soviets, as you know, have deployed ABM defenses around Moscow. They were permitted to do so under the ABM Treaty. We had started to deploy defenses in North Dakota, which is in the middle of our country and in a region over which Soviet rockets are apt to come. We started that deployment, then signed the treaty and did not deploy.

Now there is argument that SDI deployment is in violation of the treaty. Well, we still might deploy. Just put it up there. The question is whether we want to abide by the treaty more than the Soviets do. If so, we can go forward with the 100 launchers that the treaty permits; and if we do so, I suggest that they not all be of one kind. I suggest that we build ten launchers of one kind, ten serving some other purpose, ten of a third kind, and so on, so that we get the vital practical experience appropriate to our planning. Of course, we may do none of this, or we may do something entirely different. But we must get down to practical applications in addition to research. We *have* to do this if we are serious about defending this country—and we must be.

Q. Do you think it is possible that SDI will be abandoned as a chip in arms-control negotiations?

A. We have made many mistakes in the past. I hope that mistake will not be made. . . .

Q. Some people suggest that since the major proposed SDI systems are unlikely to be 100 percent effective, we should not bother to deploy. What is good about a limited defense?

A. I am very glad you brought that up. It reminds me of a little bit of interesting history.

In World War II, during the blitz, there was unanimous sentiment in favor of air defense. And, when the Soviets turned out to have exploded a nuclear device in 1949, there was again unanimous opinion, not in the least partisan, that we should have air defense. Then, a few years later, it became obvious that air defense was not enough, that we had to have lots of other defenses as well. And at that point our leaders decided that defense was too arduous or futile and we had better rely on retaliation.

101

Q. The so-called Mutual Assured Destruction or MAD policy?

A. Yes, and I believe that was a very real mistake. The psychology was wrong. Americans had been used to being completely secure from attack. American leaders reacted to the new development by saying, "If we can't have complete security, let's not have any." That is childish psychology any parent will understand. But it's still a mistake. You see, if you have no defense, you invite an aggressor to try to build up his first-strike capability to obliterate your potential retaliation. It sets up conditions under which the one who attacks first has the greater chance, placing those who are peaceful in a very dubious position. . . .

Q. One of the main points raised by opponents of SDI is the claim that it will exhaust the resources of the United States and provide no side benefits in technology with civilian applications. Will you comment?

A. I contradict those arguments and denounce them as absurd. One point is that, as long as we have no deployment, no one can estimate accurately how much it will cost, let alone whether it will exhaust our resources. I think we need to start to deploy in order to find out what the costs are likely to be. In other words, at least deploy partly. That is what the Russians have done.

Q. What about the issue that there are no side benefits from SDI research, and that it is swallowing all the dollars that might otherwise be available for other scientific research?

A. As for the side benefits, I have yet to see any technological development that was done to produce side benefits. The new technologies are the new applications. Let me give you one example.

Right now, we are working on an x-ray laser. We know it can be done. To what extent it will be effective, I don't know. But I am convinced that an x-ray laser can be used to get the image of the interior of a living cell—even the interior of the nucleus of a cancer cell undergoing division. The laser will kill the cell, but it will act so fast as to surprise the cell in the act of living and dividing. Today, the stories that you read about cell division are primarily theoretical. Nobody has seen it happen. To get that kind of insight into biology—actually see how the nucleus of a cancerous cell differs from that of a noncancerous cell— could have enormous consequences.

Q. What is this semi-feudal prejudice against new technologies that are not tightly controlled, preplanned, and developed to order?

A. All of these reactionary people pretend to be "democrats." But they do not want to hear about any new technology unless it is for their own preplanned purpose. And, if a technology also has a military purpose, then they can't bear even to look at it.

Q. Oh, indeed. And they make moral judgments about it and call it "evil."

A. They do not realize that tools themselves are not evil. The purpose of a person using a tool—an ax, a computer, a laser—may be evil; but tools themselves are not evil.

Q. In your judgment, Dr. Teller, are the Soviets ahead of the United States with respect to exotic technologies—particle beam, microwave, and laser weapons—for space-based defense systems?

A. According to my guess, the Soviets are stupid. They have worked on these weapons technologies for 25 years while we marked time, yet now they are only ten years ahead of us. If they were clever, they would by now be so far ahead of us that we would not have a chance.

Q. One of the SDI-related projects, estimated to cost $1 billion, involves creation of a sophisticated, computerized simulator called the National Test Bed. Do you expect side benefits from that?

A. I cannot be too specific. But the test establishment really has a great deal to do with command, control, and communications aspects, simulations of threat and engagement conditions utilizing various sensors and weapons, and things of that nature. In some applications, I think, it probably will have side benefits. But, as you know, it will not be completed for another five or six years, so it is difficult to project precisely what those benefits will be after it is in operation.

Q. Do you think there will be a relationship between these projects and the superconducting supercollider or SSC proposed to enable physicists to advance much further in understanding the fundamental particles of the universe and the conditions that existed in the infinitesimal fraction of a second after creation?

A. I cannot make any connections; but I certainly cannot say there are none. As is very well know, the science of physics underlies such applications as atomic weaponry, research into lasers, particle beams, and many other applications. I simply cannot predict what advances in knowledge will come.

Q. I mentioned the SSC, for which total costs of construction, computers, detectors, test equipment, and so forth are estimated at $4.4 billion, as another example of the high cost of technology, and of the fact that some of the SDI opponents are saying that the high cost of SDI will block funding of the SSC.

A. I think advanced technology will produce yet more advanced technology, but that bureaucracy will produce only more of the same. The SSC depends on technology. It is certainly possible that if we spend too much on bureaucracy that will have a harmful effect. But on developing new technologies, whether it is defense or SSC, I would not be opposed to it on the basis that it costs some money. As for SDI, we are now spending little more than one percent of our military expenditures on it. Obviously that is not enough to preempt any other program.

Knowing that the Soviets are spending much more on strategic defense, and are protesting so vehemently about us spending anything on it at all, makes me suspect that we might be spending much too little.

Q. What is the one single and overwhelming argument that you would advance as to why the United States should vigorously and immediately pursue SDI?

A. It is that we have neglected the defense of our citizens for a quarter century or more—for almost thirty years. There is a clear need to correct an obvious mistake before we are scarred forever. President Reagan has even proposed to share this defense, yet Andropov, and now Gorbachev, opposed it. That our public officials should be on the side of Gorbachev in this matter is remarkable!

Q. Why is it that all our allies—from Canada to Europe, Japan, and Australia—are divided over the concept of strategic defense?

A. Indeed, we are not unique. The same political division exists all through the Free World, with one exception, and that is Israel. In Israel, the threat of the Soviet rockets in Syria is so close, and the bitter experience so recent, that both liberals and conservatives are all for defense. They have given the obvious answer to an obvious threat; and it is peculiar that we cannot make the same step of understanding. The threat to the survival of the people is the same whether the rockets are launched from fifty miles away or 5,000 miles away.

SDI AND SCIENTISTS

SDI IS
A SCIENTIFIC FANTASY

Charles Schwartz

Charles Schwartz is a professor of physics at the University of California at Berkeley, the sponsoring institution of the Lawrence Livermore and Los Alamos nuclear weapons laboratories. Schwartz has been active for years in raising questions about the complicity of scientists with the military and the nuclear arms race. Most recently, he has been an outspoken opponent in the science community of the Star Wars program. Schwartz was interviewed in this reading by Danny Collum, a correspondent for Sojourners. Sojourners *is an independent Christian monthly magazine.*

Points to Consider:

1. Describe what happens at the Lawrence Livermore laboratory in California.
2. Why are many scientists opposed to SDI?
3. How does the protest against SDI differ from scientists' support of a nuclear weapons freeze?
4. What personal choices did Schwartz make to protest military work and the arms race?

Danny Collum, "Defying the Law of Physics," *Sojourners,* May 1987, pp. 22-24. Reprinted with permission from *Sojourners,* Box 29272, Washington, D.C. 20017.

It was easily understood by many in the scientific community, and indeed many in the public, that the ambition for providing a complete shield that would, as President Reagan said, "make nuclear weapons impotent and obsolete," was really a fantasy.

Sojourners: **What happens at the Lawrence Livermore laboratory in California? Is there a connection between Livermore and the University of California?**

Charles Schwartz: There are two central laboratories in this country that have carried the complete responsibility for research and development of nuclear weapons. One is the Los Alamos lab in New Mexico, where the first atomic bomb was developed. The second, called the Lawrence Livermore lab, is located 40 miles from Berkeley. It was established as a second bomb lab by Edward Teller, scientist and early advocate of the hydrogen bomb, and others in 1952.

The regents of the University of California hold the contract for the so-called management of both of these laboratories. That strange arrangement has been in existence since World War II. It has been the topic of sporadic complaints and protests from within the campus. But the university has maintained a rigid patronage relationship with these two leading nuclear weapons labs.

The laboratories—Livermore and Los Alamos—are enormous scientific laboratories; their combined budgets are something like $1.6 billion a year. They are highly autonomous organizations. They not only do the technical work related to nuclear weapons, but their top officials are very significantly involved in helping to shape government policy regarding all aspects of the nuclear arms race. The connection to the University of California is that the university lends its name as a kind of intellectual legitimation for the whole enterprise of nuclear weapons.

This legitimation by the university also adds to the particular credibility of laboratory officials when they go to Washington—as they frequently do, and mostly in secret—to "advise", or, I would say, to lobby for a variety of weapons programs.

From the evidence in the public record, the advice these lab officials give the government shows a very strong bias toward the continuing of the arms race. They've voiced opposition to a variety of arms control measures and have advocated varieties of new weapons. Often they think up new weapons and then go actively selling them to the Pentagon.

There has been a fair amount of publicity regarding the Lawrence Livermore lab and the Star Wars program. Some accounts have

Livermore officials initiating the program in the way you've just described. What has been Livermore's role?

In the late '60s, there was a big push for a missile defense system, the so-called ABM (Anti-Ballistic Missile) system. It was a controversial issue then. At that time in many sectors of what is called the "defense community," and certainly at the Livermore lab, scientists were very much in favor of a defense system and promoted the idea. The idea of missile defense was stopped at that time by the ABM treaty. One of the main drawbacks to the ABM was that it just wasn't likely to work. The technology wasn't there.

But research and development work continued at Livermore and elsewhere in the decade or two since then. Some people, especially Edward Teller and others in the weapons-planning business, have always dreamed of having a combination of effective first-strike offensive weapons—which we do have and are continuing to build—together with a defensive system, in order to make nuclear war a credible military tool.

So, apparently they continued research at modest funding levels through the '70s. And according to various reports, Edward Teller, perhaps with some of his colleagues, was able to persuade President Reagan that they had wonderful new technologies on the horizon that would finally make such a missile defense system possible. Apparently this had a lot to do with Reagan's coming forward with his surprise announcement of the SDI, or Star Wars program, in March 1983.

At the same time that one sector of the scientific community played a key role in initiating the Star Wars plan, other scientists have been at the forefront of opposition to the program. Why are so many scientists against SDI?

There are two reasons. First, a few scientists who took the lead have a great deal of prestige in the scientific community and considerable

Cartoon by Steve Sack. Reprinted by permission of *Star Tribune, Newspaper of the Twin Cities.*

reputation and experience from working on the inside with the government in earlier atomic bomb programs. So they have quite high credibility. These include Hans Bethe, Richard Garwin, and others with the Union of Concerned Scientists.

These people are acknowledged weapons experts but are free of the political constraints that exist for people within the Reagan administration or at the lab. In the past they have criticized various weapons systems, and they undertook a vigorous technical criticism of the SDI program when it first appeared. This provided a technical basis for the credibility of scientific criticism.

As a result it was easily understood by many in the scientific community, and indeed many in the public, that the ambition for providing a complete shield that would, as President Reagan said, "make nuclear weapons impotent and obsolete," was really a fantasy. To be able to stop thousands of nuclear weapons, each one having perhaps many decoys, and identify, locate, and destroy all of these targets in a very short time, with something very close to if not equal to 100 percent efficiency, was simply beyond reasonable technical expectations.

This early criticism was widespread within the scientific community and was accepted by many as showing that the Reagan proposal was rather nonsensical. So there was an element of technical absurdity that scientists professionally felt they ought to speak out about.

The second aspect of SDI that deserved criticism—and to my mind it is certainly more important—is that by trying to develop such a weapons system and perhaps partially succeeding in having the ability to stop some Soviet missiles headed toward us, one has created a very dangerous military posture. The combination of a large offensive nuclear arsenal with a modest defense creates the possibility for a first strike. You could knock out most of the opponent's weapons and then be able to defend yourself against a modest retaliation.

This is a posture that causes fright in the Pentagon when they imagine the Soviets developing it. And, similarly, our plans cause the Soviets great anxiety. That anxiety, of course, creates more weapons, more instability, and in fact makes the possibility of a nuclear war more and more likely. This is a prescription for a destabilizing configuration of offensive plus defensive weapons.

You mentioned that some prominent scientists have opposed previous weapons systems. But the current campaign of scientists actually boycotting SDI funding and research is something new, isn't it?

I think it is novel. Certainly there have been statements and petitions by significant numbers of scientists on various arms-related issues. For example, a nuclear weapons freeze was publicly supported by a large number of scientists in the usual way of putting their names on a statement that might be published somewhere in a newspaper.

But this SDI petition that has circulated in the last two years is more significant, because many academic scientists are pledging that they will not accept funds to work on the SDI project. For academics to be refusing money is very significant. It is a move of active resistance on the part of large numbers of scientists—resistance to going along with this government program.

There's a very familiar attitude among many academics who take Defense Department money and say, "But it's just pure science." They claim it's not a weapon they're working on. Academics and other scientists have found various ways to rationalize their acceptance of some relationship with military while still professing their opposition to certain military programs and saying that their work is not really contributing to them.

So the SDI boycott stand really is a significant one. Here we see scientists refusing to participate in the program and refusing to take the very generous buckets of money that are being offered.

Do you think that is the result of a higher political consciousness growing out of the freeze campaign and other initiatives of recent years? Or is it the perception that the Star Wars program poses a much higher, historic order of danger and thus requires more active opposition?

I'm sure there are elements of a historically higher political consciousness. Also there is the sharpness and clarity of this issue being really technically absurd. As most scientists are opposed to things like astrology, or doctrines that they feel are professionally offensive, similarly Reagan's proposal of the SDI was a professional insult that many felt obliged to respond to.

But I think the more serious consideration is its real impact on the question of peace and war. Many scientists understand that SDI would be destabilizing, making the nation and the world less secure rather than more secure. That's a more subtle judgment and gets into realms where scientists would say they are not competent. But that's an important judgment that scientists can and should engage in. And in this case, many did.

In the course of the last year, you have taken a particular stand with regard to what you will do and not do as a physicist.

Yes. My own concern and opposition to militarism and the arms race have been rather stronger than many of my colleagues. Over the last several years, I have noted two phenomena particularly. One is the increasing danger of the nuclear arms race. We have seen the development of first-strike weapons like the MX and Trident II missiles and battle-management systems that seem to be oriented toward the possibility, if not the plan, of our government to engage in an active, vigorous nuclear war-fighting situation.

The SDI caps all that off. It makes a war-fighting scenario rather more plausible and even likely because of the possibility of defending oneself against an opponent's weakened retaliation. So the heightened objective danger of the arms race has caused me to reconsider what can be done and what my own role as a physicist and physics teacher might be in response to that.

The second feature I've noted is the increasing number of reports from physics students leaving the university who say it seems extremely difficult, if not almost impossible, to find jobs that are not related to weapons programs in one way or another. The fact that an enormous portion of the physicists we train in the universities end up working on weapons is not new. But under the budget priorities of the Reagan administration, there's been a very sharp shift so that there are very few jobs—and some say none—available for people in physics and certain other technical fields that would use their trained skills in other than weapons-related programs.

For many students this is an important moral and political question. They cannot avoid making a choice when they're looking for a job. In light of that, I felt it was important to re-examine my own role and reassess my own choices and activities and complicities.

Even though many years ago I had given up various areas of research that had some relation to military work, I actually looked at the fundamental job that I still do, which is teaching basic physics at the introductory freshman level, the advanced undergraduate level, and the graduate level. And I was forced to acknowledge the obvious fact that the people who make weapons get their basic training in colleges and universities from people like me, perhaps many of them from me in particular.

This basic-level physics is used as a foundation for every physicist, mathematician, chemist, or engineer to build upon their technical skills. When there are not many alternative jobs for these people other than weapons work, it becomes harder to justify the teaching work by saying, as most people do, "Well, teaching is just pure. It's just the knowledge of physics. It could be used for good or for ill. So let's just hope that it's used more for good than for ill."

I had to face, and finally accept, the reality that what I'm doing in continuing as a physics teacher is performing a vital service for the Pentagon. I was providing essential raw material for the arms race by training the students who will be the next generation of weapons designers. And I think that perhaps they may be the last generation of weapons designers if this march toward nuclear war that I see happening is not halted and reversed.

This caused me to make a choice, and I chose to inform the chair of my department that I was no longer willing to teach most of the regular physics courses. I said this was because I no longer wanted to be complicit in what appeared to me clearly to be preparations for nuclear war—something that would have to be called unconscionable.

I said that as a teacher I felt that I had to withdraw my services. I even asked to be considered "a kind of conscientious objector" but still within the profession of physicist and physics professor.

How has that worked out practically?

Not too badly. I was a little concerned. It was not my objective to confront my department chair and my university. Ultimately they are not responsible for the problem.

The chair first responded by asking if I planned to resign as a professor. And I said no, it's not that simple. Then I asked if he would simply assign me to courses teaching physics to liberal arts majors and to people going into the biological sciences, where they were rather unlikely to use the knowledge I taught them for weapons work.

I also teach a course on the arms race and another course on science and politics. I felt that these were positive uses of my knowledge and skills as a teacher and that I certainly wanted to continue doing this and hoped that the university and the department would be able to accommodate me.

111

So, at least for the present, I have been given those teaching assignments, and it has worked out. I don't know if, at a later time, I may be ordered to teach some course that I don't find morally acceptable. I hope we won't have that kind of confrontation. But we'll see.

SDI AND SCIENTISTS

SDI IS
A SCIENTIFIC REALITY

Robert Jastrow

Dr. Robert Jastrow founded and was for 20 years director of the God-dard Institute for Space Studies at NASA. Dr. Jastrow was the first chairman of NASA's Lunar Exploration Committee, which set the scien-tific goals for the exploration of the moon, and he was awarded the NASA Medal for Exceptional Scientific Achievement. Dr. Jastrow is also founder of the George C. Marshall Institute, the Washington-based organization that develops educational programs relating to fields of science with the goal of affecting public policy. He is interviewed in this reading by Cynthia V. Ward, a senior editor for Conservative Digest.

Points to Consider:

1. Dr. Jastrow says that SDI will be *"100 percent* effective in protect-ing the American people from the danger of a first strike." How will this be possible?
2. Describe the technological developments that have been made in space-defense research.
3. What countermeasures could the Soviets use against SDI?
4. Compare and contrast U.S. and Soviet technological development of space-based defenses.

Cynthia V. Ward, "Dr. Robert Jastrow Tells How Strategic Defense Will Work," *Conservative Digest,* May 1987, pp. 37-48. Reprinted with per-mission of *Conservative Digest.*

Our country can have by the mid-1990s a defense that will utterly paralyze any Soviet effort to launch a first strike against the United States.

Q. Dr. Jastrow, what exactly could we deploy in space-defense technology by the mid-1990s? And how would it work?

A. As the George C. Marshall Institute explains in its report, "Missile Defense In The 1990s," the initial defensive system would consist of two or three layers—boost phase (that is, those weapons designed to destroy Soviet missiles as they arise from their silos); late midcourse (those weapons which shoot down Russian missiles that survived the boost-phase defense and are still en route to U.S. soil); and, terminal phase (those weapons based on American land which would destroy any remaining Soviet missiles as they come in for the attack).

All these would use the "kinetic-energy" systems, which are modified versions of the heat-seeking air-defense missile. This type of weapon senses its target by the heat it emits, homes in, and then destroys the target by force of impact. . . .

The technology for this system has been tested several times in space and has worked perfectly every time. Last September, for example, it was tested in an experiment in which two spacecraft maneuvered around another. One of them, the Strategic Defense Initiative (SDI) defense vehicle, then zeroed in on the other, a simulated Soviet missile, and scored a *bulls-eye collision* using this air-defense technology, which was actually just a modified air-to-air missile. Indeed, the defensive weapon hit within twelve inches of the intended target point, and of course destroyed the target, because at this high speed the force of the collision released ten times as much energy per pound as a TNT explosion. . . .

Our country can have by the mid-1990s—starting in 1994 if the system is approved this year—a defense that will utterly paralyze any Soviet effort to launch a first strike against the United States. It's a defense that we estimate to be 90 percent effective. But it will be *100 percent* effective in protecting the American people from the danger of a first strike. . . .

Q. It's a perfect deterrent?

A. Yes. Because a Soviet general who is planning a first strike must count on wiping out our forces of retaliation; otherwise his own homeland will be in ruins within sixty minutes. And if he knows that 9/10ths of his warheads are going to be shot down—that only one in ten is going to get through—and he doesn't know *which* one that will be, he cannot plan a strike against our command centers, our sub-

114

marine bases, or our critical nodes. He will not launch an attack under those circumstances.

The Union of Concerned Scientists says that the one bomb in ten that does get through will destroy American cities. But notice the fallacy in their reasoning. According to the CIA, the Soviet Union will have at least 10,000 nuclear warheads by the mid-1990s. If we shoot down nine out of ten of those, and the one in ten that gets through lands on a city, the implication is that all 10,000 were targeted on American cities. Now why would the Soviet Union spend a trillion dollars on building a 10,000-warhead arsenal, most of which are capable of landing within 300 yards of the target—that is, they are designed to knock out hardened military targets—and then blow this whole arsenal on targeting it against American cities, the one act which is guaranteed to unleash our own forces and be suicidal?

Q. Some critics of SDI say that there are less-expensive ways of enhancing our nuclear deterrent, such as installing superhardened missile silos or moving more of our missiles out to sea.

A. Now those are honest suggestions. But if you're really asking "How can we protect the American people against the Soviet threat at minimum cost?" I think the answer is that if you keep improving your nuclear weapons and retaliatory capability the arms spiral will only continue to escalate. It's not a stable way to go. The total number of nuclear warheads in the arsenals of the two superpowers in 1972, when Nixon and Brezhnev signed the Anti-ballistic Missile (ABM) Treaty, was 4,800. Today that number is 16,000. Furthermore, it is acknowledged by virtually everyone that the Soviets are cheating on the ABM Treaty.

The SDI offers another avenue. And you may say well, that sounds pretty good, but how much does it cost? And the answer given by certain people, such as ex-Secretary of Defense Schlesinger, is a trillion dollars. That number was pulled out of a hat.

115

CHAPTER 78: Darth Vader suddenly wants to talk...

CLICK CLICK CLICK

GENEVA

OLIPHANT

Reprinted by permission: Tribune Media Services.

When we look at the updated air-to-air missile technologies we find that the cost of a system which would give us this 90 percent protection—full protection against a first strike—is $121 billion. Other estimates vary, but the figure is surely in the neighborhood of $100 billion.

If you think that still seems like a lot of money, I must tell you that this $100 billion would be spread out over ten years of development and then deployment. So the cost would run $10 to $15 billion every year, approximately 20 percent of the current American budget for strategic offensive weapons. In fact, $10 billion a year is only *three percent* of the defense budget. For three percent we get guaranteed protection from a nuclear first strike.

One more thing about cost. The ATF—our new fighter—cost $65 billion. All of the major strategic weapons programs cost in the same ballpark as this strategic defense we are talking about....

Q. It has been almost exactly four years since President Reagan proposed the SDI. Since that time, what specific technological developments have been made in space-defense research?

A. The product of these years has been a lot of progress on lasers, which now tells us which types are the most promising but still indicates that using them in American defense is an end-of-the-century matter.

116

The Marshall Institute panel that produced our report thought lasers were at least three or four years behind the kinetic-energy weapon. But we know we must still invest in laser research, because if we only build the kinetic-energy defense, and then leave it for say fifty years, eventually the Soviets will figure out how to overcome it and we will have to have lasers waiting in the wings. . . .

Q. What are the most threatening countermeasures that the Soviets could employ against the SDI?

A. The most significant countermeasure is the decoy. Two other possible countermeasures, which are much talked about, are potentially serious.

The fast-burn booster is a rocket that springs upward and burns out so quickly that our boost-phase kinetic-energy weapons would not have time to reach it. The troubles with this countermeasure are cost and design.

When a missile rises in the air it's being pushed from below and it's like a pencil balanced on your fingertip. It tends to fall over. As soon as it starts to fall, the sideways pressure of the air rushing past pushes it farther. The rocket designer counters this by mounting the rocket nozzle so it can swivel in different directions. So as soon as the pencil starts to fall to the right, the rocket nozzle swivels to the right, deflects the rocket exhaust and that corrects the rocket's direction. This is why rockets fly straight.

If you have a fast-burn booster it moves up through the air faster, and the pressure of the air going past it is greater. So it exacerbates this problem of making the missile fly straight, because when it starts to topple sideways the force is greater. Research has found that the forces that can be applied by swiveling the rocket nozzle are not adequate to keep a fast-burn rocket flying straight. So in principle you can't build a fast-burn intercontinental ballistic missile (ICBM).

Now maybe you can figure out how to do this if you work at it hard enough. But while you might be able to build a fast-burn *small* ICBM like the Midgetman, it will be impossible to build a fast-burn SS-18. The SS-18 is as high as a ten-story building and weighs 200 tons. The inertia is so great that the toppling tendency could never be countered by any known system of rocket engineering.

The consequence is that if the Soviet Union decides to build a fast-burn arsenal and supplement or throw away its present arsenal—that's a minimum 10,000 warheads in the 1990s—they'll have to build a fast-burn Midgetman arsenal. Now, the Midgetman is extremely expensive, costing about $100 million *per warhead*. This means that a fast-burn Midgetman arsenal in the Soviet Union would cost them a cool trillion dollars. All of this just to counter a defense which we calculate to cost $121 billion.

Space mines have also been suggested as a threat to the sensor satellites. It is proposed that the adversary can deploy the space mines covertly to follow our satellites, ready to close in and destroy them at the start of the enemy attack.

In the Marshall panel's judgment, however, space mines are not a useful mode of attack against a sensor satellite. They cannot be deployed covertly because the sensor satellite network, which is equipped to observe all missile launches, will see the rockets that place the space mines in orbit. Subsequent maneuvers of the mines in orbit are also likely to be observed, either by the sensor satellites themselves or by the Space Track radars currently operated by NORAD. If a space mine that has kept a respectful distance from a U.S. satellite starts to move in toward the satellite, that fact will be observed.

Moreover, the sensor satellites are equipped by the nature of their instrumentation to detect any space mines following them at a close distance. The space mine must fire small rockets periodically to compensate for differences in drag and stay close to its quarry. When it does so, it is likely to become visible to one of the sensor satellites. It should be noted that if the space mine is deployed covertly by the adversary, it can be taken out covertly by the U.S. Our panel concluded that space mines are not among the more serious threats to survivability of space-based defenses.

Q. Is it logical to take the view, as some of the critics do, that if we did deploy the SDI the Soviets would then devote more time and money towards overcoming some of these problems with countermeasures; towards developing better ASAT weapons or something? In other words, that they would escalate the "arms race in space"?

A. The response which has been suggested by a number of Congressmen and others is that the Soviets might build up their offensive nuclear arsenal to "overwhelm" our defense. But if we have a space-based defense, that doesn't stand up to scrutiny at all. For example, suppose we have a defense which is not 90 percent effective as we estimated, but only 80 percent effective. To overwhelm that defense means building up their arsenal to get as many warheads through the defense as they would have without it.

Against an 80 percent defense, that means increasing the Soviet arsenal by a factor of five. Now you're talking about spending $5 trillion to build a 50,000 warhead arsenal. But when you delve deeper into the analysis it turns out even that is not enough. An 80 percent defense means on the *average* we'd shoot down four out of five of the other fellow's warheads. But in any particular case we may shoot down *more* than four out of five. The point is this: When you flip a coin, on the average you get heads half the time. But you might flip it and get tails four times in a row. And if you do a statistical analysis to find out how

many warheads the Soviet Union must expend in order to be 90 percent confident that they're going to get *one* warhead through our
defense, the answer turns out to be not five to one but roughly ten to
one. So they wind up having to build and use 100,000 warheads to
maintain a credible first-strike threat. That's *$10 trillion*. And even then
they have a 10 percent chance of failure—a failure that would be
catastrophic for them because we would then be able massively to
retaliate.

**Q. Overall, where is the U.S. compared to the Soviets in terms
of technological development of space-based defenses?**

A. The Soviets are well ahead of us on lasers, especially laser ASATs,
which they have been testing. They're well ahead of us on deploying
a nationwide ABM defense using old technology, with a small nuclear
weapon on the missile instead of a smart bullet. We're supposed to
be somewhat ahead of them on the computer and the smart bullet and
the kinetic-energy technology. But they're scrambling to catch up, and
they've been spending not $3 or $4 billion a year as we have for the
past three years. They've been spending between $10 billion and $15
billion a year every year for the past ten years on this program. They've
spent more than $100 billion on it so far. And whatever they can't figure
out using their own limited computer expertise, they will certainly try
to steal from us, as they've done in the past. So I can't see them lag-
ging behind us on smart bullets much longer.

**Q. How much time does the U.S. have to get going? Will there
be a point beyond which, if we keep stalling on SDI, we won't
be able to catch up with the Soviets?**

A. I think the critical year will be 1995, because around that time the
Soviets will have completed deployment of their fifth-generation missiles,
including mobile missiles that can't be verified in any arms-control agreement. According to the CIA, they could have as many as 12,000 to
16,000 ICBM warheads, and as many as 21,000 nuclear weapons at
that time. By the early 1990s their network of large battle-management
ABM radars will be completed. The last three were just started. They
have production lines for turning out ABM computers in quantity. So
by the mid-1990s they will have a combination of a massive first-strike
arsenal and an ABM defense.

Now, that ABM defense may not be good enough to counter a first
strike from the United States. But the problem is not to keep *us* from
launching a first strike. Their ABM defense will be *very* effective against
the ragged retaliatory counter-strike that would get off the ground after
they've wiped out 95 percent of our ICBMs, two-thirds of our bombers
and cruise missiles, and half our submarines. Mind you, the only deterrent we now have that's really worth anything is our Trident submarines.
There will be just six of them on station—not in port—at any given time

in the 1990s. The whole safety of the United States rests on that thread. And the Soviets have well over 100 attack submarines. That's sixteen submarines that could be assigned to every Trident to follow it everywhere and keep it in a cage. There's no chance we can escape from them.

So this is the situation we must be ready to meet around 1995. If we don't get the necessary funding for SDI, Soviet ABM deployment will probably make an impact around the beginning of the next Administration, because the Russians can't conceal what they're doing from anyone much longer. Then the question is, will there be catch-up time? When Pearl Harbor hit us, things were slow-paced. We had a couple of years to gear up. Today we'd have only twenty minutes. We'd better start thinking very hard about that fact.

RECOGNIZING AUTHOR'S POINT OF VIEW

This activity may be used as an individualized study guide for students in libraries and resource centers or as a discussion catalyst in small group and classroom discussions.

The capacity to recognize an author's point of view is an essential reading skill. Many readers do not make clear distinctions between descriptive articles that relate factual information and articles that express a point of view. Think about the readings in Chapter Three. Are these readings essentially descriptive articles that relate factual information or articles that attempt to persuade through editorial commentary and analysis?

Guidelines

1. Read through the following source descriptions. Choose one of the source descriptions that best describes each reading in Chapter Three.

Source Descriptions
   a. Essentially an article that relates factual information
   b. Essentially an article that expresses editorial points of view
   c. Both of the above
   d. None of the above

2. After careful consideration, choose one reading that you agree with the most. Be prepared to explain the reasons for your choice in a general class discussion.

3. Choose one of the source descriptions above that best describes the other readings in this book.

CHAPTER 4

STAR WARS
AND MORAL CHOICES

17

STAR WARS
AND MORAL CHOICES

MORAL RHETORIC
AND CONFUSION IN THE
STAR WARS DEBATE:
AN OVERVIEW

Edward T. Linenthal

Dr. Edward T. Linenthal wrote the following reading for The Christian
Century *in his capacity as associate professor of religion and American
culture at the University of Wisconsin-Oshkosh. In 1986-1987 he was
a postdoctoral fellow in the Defense and Arms Control Studies Program at MIT. He is finishing a book on SDI and American culture.*

Points to Consider:

1. Why have both proponents and opponents of SDI argued their case
   on moral grounds?
2. Describe the arguments that SDI enthusiasts use to promote missile
   defense.
3. Summarize the reasons why opponents do not find SDI morally
   compelling.
4. Why do SDI advocates believe there are no substantive moral objections to missile defense?

Both proponents and opponents of SDI have argued their case on moral grounds—whether they emphasize the morality of motive or the morality of consequence. . . .Thus, the debate over SDI has crystallized, sometimes in ironic ways, America's struggle to comprehend the moral significance of nuclear weapons.

One of the "image achievements" (to use political analyst Kevin Phillips's term) of President Reagan's "Star Wars" missile defense program has been its appeal on moral grounds. For proponents of the Strategic Defense Initiative (SDI), the proposed space-based defense system represents the ultimate moral triumph: it will release us from the seemingly stubborn dilemmas of the nuclear age, rendering nuclear weapons "impotent and obsolete."

Of course, both proponents and opponents of SDI have recognized that public opinion on the issue will be formed by moral as well as technological and strategic arguments. Both proponents and opponents have argued their case on moral grounds—whether they emphasize the morality of motive or the morality of consequence. Some have also used scriptural references to support their respective positions. Thus, the debate over SDI has crystallized, sometimes in ironic ways, America's struggle to comprehend the moral significance of nuclear weapons.

SDI: A New Way of Thinking

Shortly after Reagan endorsed SDI on March 23, 1983, SDI enthusiasts began marketing the president's vision as a Reaganite response to the "new ways of thinking" that Albert Einstein had called for in the early years of the nuclear era. Like the impassioned critiques of deterrence that have led antinuclear activists to call for a nuclear freeze or even more radical plans for disarmament, SDI could be perceived and marketed as an alternative to the "delicate balance of terror" of Mutual Assured Destruction (MAD). Writing in the *Air University Review,* Barry Smernoff used the language of the antinuclear movement to characterize SDI proponents as "new abolitionists," who "preach against the immorality of nuclear deterrence and nuclear war" ("Images of the Nuclear Future," May/June 1983). The simple and simplistic slogan, "SDI will destroy weapons, not people," captured the appeal of this "new" way of thinking. (An early attempt to codify such sentiment was Colorado Republican Representative Ken Kramer's "People Protection Act" of May 1983, which advocated "people protection" through missile defense.)

SDI enthusiasts have argued that MAD, rather than being an inevitable part of the nuclear age, is an immoral doctrine, since it threatens civilians.

Cartoon by David Seavey. Copyright 1983, *USA Today*. Reprinted with permission.

High Frontier, an influential advocacy group for missile defense, says that technological advances offer the United States a "historic, but fleeting, opportunity to take. . .destiny into its own hands" ("Proposed Statement of U.S. Policy," 1982). Missile defense would, at long last, allow us to escape from the debilitating fear of nuclear vulnerability. The president has himself argued this issue bluntly to a radio audience: "Isn't it time to put our survival back under our own control?" (*Boston Globe,* July 13, 1986).

Secular and religious advocates of SDI have claimed that missile defense would allow America to return to "moral" principles of defense;

it would, according to then Secretary of Defense Caspar Weinberger, express the "best characteristics of democratic principles" ("Ethics and Public Policy," *The Fletcher Forum,* Winter 1986). According to Lewis Lehrman, SDI would restore to the president the ability to carry out his oath to "preserve, protect and defend the U.S." ("A Moral Case for 'Star Wars,'" *New York Times,* February 19, 1985). Progress in SDI has been viewed, consequently, not only as a strategic imperative but, as Weinberger exclaimed, "a visionary moral quest" ("Strategic Defense Initiative: Creating Options for a Safer World," *Los Angeles Times,* July 9, 1985).

SDI: The Only "Moral Strategic Nuclear Military Policy"

Certain groups have even claimed that missile defense will restore Christian principles to the nation's defense policy. The Coalition for the Strategic Defense Initiative announced that over 2,000 ministers, priests, and rabbis have signed its "Clergy Statement," which declares that "no nation has the right to renounce its duty to defend its people against unjust aggression." SDI, by posing "no threat to human life," is "morally obligatory for the American people and their government." The Religious Coalition for a Moral Defense Policy, a coalition of conservative Protestant, Catholic, and Jewish groups, issued a statement in support of SDI on February 4, 1986, calling it the only "moral strategic nuclear military policy." The National Association of Evangelicals' program "Peace, Freedom, and Security Studies" declared that SDI could not only restore a moral defense policy but "could provide the occasion for new efforts at mutual security arrangements."

Pro-SDI Catholics interpreted SDI as a direct response to the Catholic bishops' challenge to the nuclear status quo in their celebrated pastoral letter, *The Challenge of Peace, God's Promise and Our Response.* Hans Mark, former technological adviser to the president and later chancellor of the University of Texas, believes that the bishops' criticism of deterrence was a motivating factor behind Reagan's "grand vision" speech. The president may also have been influenced by a Catholic deeply interested in the issues raised by the bishops, Admiral James D. Watkins, Navy chief of staff. In a crucial prespeech meeting on February 11, 1983, Watkins argued for missile defense as a "moral imperative." The president's sensitivity to the bishops' critique revealed, according to Mark, that "Reagan is really a soft touch . . . he's not a tough guy like Harry Truman was" (*National Catholic Reporter,* May 10, 1985). Writing in *This World,* Joseph Martino argued that since SDI raises no problem of immoral intentions, it answers the bishops' call for a way to move beyond the "perverse logic" of deterrence ("'Star Wars'—Technology's New Challenge to Moralists," Fall 1984). Likewise, in *Catholicism and Crisis,* Kenneth Kemp argued for SDI as the "third road" to peace. "SDI," he wrote, "more adequately reflects the values and vision of the kingdom of God" than does either disarmament or deterrence ("The

Moral Case for the Strategic Defense Initiative," June 1985). Both George Weigel's book *Tranquillitas Ordinis* and Philip Lawler's *The Ultimate Weapons* express pro-SDI Catholic sentiment. For Lawler, especially, the moral case for SDI is so clear that "if such a system could be devised, American Catholics should support it!"

Some Christians have invoked Scripture in support of SDI. In a sermon titled "Surviving the Nuclear Age," D. James Kennedy told his parishioners of Coral Ridge Presbyterian Church in Fort Lauderdale, Florida, the story of Nehemiah rebuilding the wall around Jerusalem. Now, he said, due to the miracles of technology, the United States would soon be able to shoot down "1,000 missiles a second." Kennedy called for prayers that "the wall around America may be built again. . .that we may enter a new age where nuclear weapons are no more." The call for a fortified wall also appeared in the newsletter of Beverly La Haye's Concerned Women for America. Amid pleas to protest legalized abortion, support a revival of religion on campus, and see that Central American "freedom fighters" get supplies, the newsletter expressed the hope that America will "'rebuild the wall' of our national defense," and it offered a prayer for a "space shield. . .functioning as soon as possible" (November 1986). Both President Reagan and Thomas Moore, former executive director of the Coalition for the Strategic Defense Initiative, have drawn support for SDI from Luke 11:21: "If a strong man shall keep his house well guarded, he shall live in peace" (see Moore's letter to the *New York Times,* March 17, 1986, and Reagan's radio address in *Weekly Compilation of Presidential Documents,* July 22, 1985).

Opponents Do Not Find SDI Morally Compelling

In the face of such apparent moral clarity about the benefits of SDI, some proponents have found it hard to understand the opposition. Speaking at Tufts University, Weinberger said that he was "baffled" that the "idea of defending. . .one's notion of the good should cause such an ethical dilemma." He blamed opposition to SDI on "liberal" moral relativism that denies the importance of defending the values of the West. This, mixed with a dangerous innocence about the threat from the Soviet Union, engendered, he believed, unwarranted qualms about SDI. But, said the secretary, "we have created the freest, most prosperous and strongest nation in the history of the world. We have a moral obligation to defend it—and we will."

Opponents of SDI, perceiving the nuclear world in quite different terms, do not find SDI so morally compelling. For those who have come to the sorrowful realization that humankind has forever left the prenuclear age, SDI appears to be an infantile attempt to return to this prenuclear womb. It has also seemed another manifestation of technological arrogance. The English antinuclear activist E. P. Thompson thinks SDI is driven by the American desire to return to the "years of American superiority, 1945-1950." This "ideological delirium," said Thompson, is "attuned to all the worst traditions of American right-wing populism" (*Star Wars: Science Fiction Fantasy or Serious Possibility?* [Penguin, 1985]). The editors of *Christianity and Crisis* agreed, arguing that SDI emerged not from "love of peace, but lust for hegemony" (May 14, 1984).

Joseph Nye, director of the Center for Science and International Affairs at Harvard's Kennedy School of Government, contends that the moral arguments in favor of SDI display "stunted moral reasoning." SDI, Nye argues, is "no more a moral imperative than are alternative ways of enhancing deterrence." Finally, the morality of such an initiative will "depend on the consequences, not the motives" (*Los Angeles Times,* June 5, 1986).

Church Groups Have Begun to Criticize SDI

A growing number of church groups have begun to criticize SDI, relying not on a distinctive theological critique of the program but on secular arguments. In 1984 the Unitarian Universalists passed a resolution to "Stop Space Weapons: Resume Space Cooperation," and the Church of the Brethren's General Board offered a resolution to "keep outer space weapon-free." In 1985 the American Friends Service Committee stated, "We have no faith in Star Wars." SDI is not, it said, a moral alternative, but would lead instead to the "material and spiritual impoverishment of our people." Joining the chorus in 1985, the Reformed Church in America, the Christian Church (Disciples of Christ), the Episcopal Diocese of Washington, D.C., and the United Church of Christ all stressed that SDI would spur an arms race in space and lead to a less secure future. In 1986 the 198th General Assembly of the Presbyterian Church (U.S.A.) called for a halt to "research, development, and testing plans for space-based missile defense systems."

In the same year the United Methodist Council of Bishops, finishing a two-year study of nuclear issues, published *In Defense of Creation: The Nuclear Crisis and a Just Peace,* in which they noted "overwhelming skepticism" about SDI among scientists, and worried about SDI's offensive implications, its effect upon arms control and its cost. The Methodist bishops found support for their stance in Scripture, reminding readers how often the "Scriptures warn us against false hopes for peace and security." Drawing a similar lesson from the Scriptures was Robert Jewett, who reminded readers of The Christian Century that Isaiah had once warned against a false sense of security when the Israelites signed a mutual defense treaty with Egypt 27 centuries ago. The seductive appeal of SDI offers the United States those same "fatal illusions" (May 18, 1983). Also worried about SDI's seductive appeal were the editors of *Sojourners.* They understood Reagan's Grand Vision speech as "the first blow of a concerted effort to co-opt the peace movement . . . and steal the moral high ground" (May 1987). The journal called attention to "Peace Pentecost" Sunday, June 7, 1987, when churches across the nation planned to protest the "idolatry" of placing "hope for security in technology of our own design" (*Sojourners,* May 1987).

A number of Catholics, including several bishops, have taken issue with the claim that SDI is an appropriate response to *The Challenge of Peace.* In an impassioned argument against SDI, priest and former congressman Robert Drinan argued against rushing to alter the nature of deterrence, for "deterrence is the only reason Vatican II and now the U.S. Bishops permit at least on a temporary basis possession of nuclear weapons" ("Star Wars Leap Could Escalate Arms Race," *National Catholic Reporter,* May 6, 1983). The editors of *America* declared that the peace promised by SDI was not the "true peace" of the bishops' pastoral, for such peace could not be brought about by a "machine we can buy" (November 23, 1985). Cardinal Joseph Bernardin, the archbishop of Chicago, and Cardinal John O'Connor, the archbishop of New York, commented at length about SDI in testimony before the House Foreign Service Committee in June 1984, during a debate over the MX missile. In a joint written statement they argued that "technological decisions must be governed by political choices which in turn should be governed by a moral vision." Bernardin questioned the wisdom of SDI, noting that while the "objectives are desirable," the consequences might not be. They concluded that "from the perspective of our Pastoral Letter, we support efforts to prevent the initiation of a nuclear race on yet another frontier—outer space." Almost a year later, in a speech prepared for a conference on Religion and International Relations held at the University of Missouri-Columbia, Cardinal Bernardin reiterated his "misgivings" about SDI: "While I understand the motivation behind the SDI, I am very skeptical of its consequences on the arms race" ("Morality and Foreign Policy," March 7, 1985).

SDI: Product of American Nuclear Anxiety

SDI should be seen, in part, as the product of American culture's latest period of struggle with nuclear anxiety. Ironically, many SDI enthusiasts have begun to utilize the language of the peace movement. They have criticized deterrence as vociferously as have many in the antinuclear movement, and they have argued that SDI is the only "realistic" way to initiate "new modes of thinking." SDI advocates have argued that both the motives and the consequences of SDI are good, and that there are simply no substantive moral objections. (I personally came to realize how deeply some SDI enthusiasts feel about this point during a speaking engagement on the impact of SDI in American culture to the SDI working group of the National Security Council in June 1987. There was a palpable sense of "injured innocence" among the members of the audience, a general puzzlement over the opposition to SDI, given the—to them—self-evident moral arguments in favor of it.)

As SDI advocates have turned to arguments traditionally used by the antinuclear movement, some opponents have turned away from that same language toward a polite, if distant, embrace of deterrence, strategic stability and the importance of arms control. Some opponents have been cynical about the motives behind SDI, regarding it as a veiled attempt to checkmate the Soviet Union in space or as a way to undercut the power of the antinuclear movement. Even for those opponents willing to grant sincerity of motive, the consequences of SDI render the good intentions meaningless.

A Battle for "Image Achievement"

The battle for "image achievement" regarding the morality of SDI will continue. Republican candidates must pay at least lip service to the moral, strategic, and technical potential of missile defense. Even Democratic candidates will find it difficult to abandon SDI.

Amid such political posturing it will take a wise and courageous candidate to remind the nation that the real "visionary moral quest" will begin when the superpowers realize that the cheap grace promised by missile defense will ultimately prove as illusory as all other attempts to escape from the problems and paradoxes of the nuclear age.

18 STAR WARS AND MORAL CHOICES

THE MORAL CASE FOR THE STRATEGIC DEFENSE INITIATIVE

Kenneth W. Kemp

Kenneth W. Kemp wrote the following reading for Catholicism in Crisis *in his capacity as captain in the United States Air Force. Mr. Kemp also teaches ethics in the Department of Philosophy and Fine Arts at the United States Air Force Academy.*

Points to Consider:

1. How did we arrive at the "supreme crisis"—the problem of war in the modern world?
2. Describe nuclear disarmament. What are the dangers of disarmament?
3. Summarize the deterrence policy.
4. Why does the author believe strategic defense is morally superior to disarmament and deterrence?

Kenneth W. Kemp, "The Moral Case for the Strategic Defense Initiative," *Catholicism in Crisis,* June 1985, pp. 20-23. Reprinted with permission from *Crisis* magazine, P.O. Box 1006, Notre Dame, IN 46556.

Defense against nuclear weapons raises no problems of intending (in any sense) what it would be immoral to do.

The whole human race faces a moment of supreme crisis on its advance towards maturity.[1] So warned the Fathers of Vatican II in opening their discussion of the problem of war in the modern world. How did we get to this moment of supreme crisis?

Historical Background

The roots of the crisis can easily be traced at least as far back as Eighteenth Century France where, within a space of a few years, there occurred two events which laid the foundations for the problem we now face. The first was the Montgolfier balloon ascent, which began man's conquest of the air; the second was conscription into the Republican armies, which began the nationalization of war. The century and a half that followed these events saw a radical transformation in the nature of war. The population-as-national-resource of the French revolutionaries was soon transformed into the population-as-target. But as long as wars were fought on the ground, actual attacks on the enemy population were, by and large, first impossible (because the enemy army stood in the way) and then unnecessary (because once the army in the field was defeated, the population at home saw that there was no alternative to surrender).

The conquest of the air changed all that. Air power made it possible to attack the home population as a means to winning the war. Although the strategy was tried during World War I, technology at that early date was still too primitive and the strategy had no effect on the outcome of the war. But the horrors of the protracted trench warfare that was World War I made the development of air power look like an attractive alternative—a quick blow to the enemy homeland should end a war at relatively low cost, or so the strategists thought. It did not, of course, but the targeting of civilian populations nevertheless, and in spite of military opposition, came into its own. The technology used through most of the Second War was inefficient (it took a 1,000-plane raid which lasted three days to destroy Dresden), but effective. But World War II made its contribution to technology as well as to strategy, for the war saw the development both of the ballistic missile and of the nuclear bomb. It was the combination of these two inventions—an invulnerable delivery system and a weapon small enough to be carried, but large enough to be "useful"—in a moral context which tolerated the destruction of enemy cities that created the current predicament, the "supreme crisis" of the opening passage.

Reliance on Nuclear Weapons

It did not have to be this way, of course. Immediately after the war, the United States, which had a monopoly on the possession of nuclear weapons, offered to give them up through the Baruch Plan. But this offer was rejected by the Soviet Union. Immediately after the war, the United States demobilized the largest army the world had ever seen, but the Soviet Union, in Czechoslovakia, in Berlin, and in Korea, to cite just the most salient incidents, made it clear that, though the Nazi menace had been destroyed, the threat to the free world was not in the least diminished. More recent events, in Czechoslovakia (again), in Poland, and in Afghanistan, make it clear that the threat is still before us.

The response we chose, reliance on nuclear weapons, was an economy move—nuclear weapons were cheaper, and more convenient, than maintaining a large conventional army. But the choice we made created a new danger, for the very weapons on which we relied for our defense came to pose a danger of their own. The American Bishops, in their recent pastoral letter, summarized the situation as follows:

we perceive two dimensions of the contemporary dilemma of deterrence. One dimension is the danger of nuclear war, with its human and moral costs. . . .The other dimension is the independence and

133

freedom of nations and entire peoples, including the need to protect smaller nations from threats to their independence and integrity.[2]

This reliance on the specter of retaliation to secure basic human values is, in President Reagan's words, "a sad commentary on the human condition."[3] What can we do about it? Three general lines of policy are available, each with variations.

Nuclear Disarmament

The first is disarmament: We could get rid of nuclear weapons. Two variations of the disarmament policy are available—unilateral and multilateral. The first ignores the fact that the current situation is not just a bad one, it is a *dilemma.* The Bishops rightly insist that

The moral duty today is to prevent nuclear war from ever occurring and to protect and preserve those key values of justice, freedom, and independence which are necessary for personal dignity and national integrity.[4]

Unilateral disarmament may avoid one of the horns of the dilemma, but only at the risk of impaling us all on the second.

The second variation of the disarmament policy is multilateral disarmament. Surely if both sides agreed to give up all nuclear weapons, then deterrence could be achieved, as it once was, conventionally, by our ability to deny the enemy success on the field of battle, rather than by threat of unacceptable retaliatory losses. And if deterrence should fail, at least there would be no chance that the war which ensued would destroy the whole planet. The problem here is not over whether a nuclear-free world would be a good thing—everyone agrees that it would. The question is rather over whether we can get there, and if so, how? For there are significant, perhaps insuperable, obstacles which stand between us and a nuclear-free world, at least along any of the most commonly proposed roads to that end.

The Dangers of Disarmament

Disarmament would only provide security if we could be sure that the other side actually did disarm. And verification of arms control agreements is a tricky business. At the current relatively high levels of armament, the tolerable margin of error is within our technological capabilities—if we miscount by 10 missiles, or perhaps even by 100, it is a matter of little strategic significance. But in a disarmed—or nearly disarmed—world even such a small difference in arsenal size would give decisive strategic advantage to the nation that had it. The incentives for cheating on arms control would be great and would only be magnified by the fear that the other side might be cheating as well.

A related danger is that small arsenals, whether achieved via gradual reductions on the way to disarmament or via cheating on an agreement to disarm completely, are highly unstable. This instability arises from each nation's perception that the only way to survive the crisis is to destroy the other nation's nuclear forces in a preemptive strike. The prospect of achieving a successful pre-emptive strike on a nation with a strategic arsenal of a thousand warheads, properly deployed, is fairly small. There is no point in attacking unless one can assure that the victim of the pre-emptive strike has no significant retaliatory capability, and gaining that kind of assurance is no easy matter. But the probability of success in destroying an arsenal of only 10 weapons is much greater. That is why small nuclear arsenals are relatively unstable. The Bishops rightly emphasize the importance of considering stability when discussing the criteria to be used in evaluating new weapons systems.[5]

Complete nuclear disarmament, then, makes cheating too rewarding; even approaching it is dangerously destabilizing.

The Deterrence Policy

A second policy alternative is to stick with the current strategy of deterrence by threat of nuclear retaliation, or some variation of it. The variations on this policy pertain to the nature of the retaliatory threat. The history of the U.S. nuclear policy seems to be a rather uneasy vacillation between general threats to destroy Soviet society and more focused threats on particular aspects of Soviet power (pre-eminently, but not exclusively, on its military forces). This strategy of deterrence by threat of punishment has kept "peace of a sort" for some forty years now, but it is, for numerous reasons, less than ideal.

First, execution of certain threats would clearly be immoral. Any act of war aimed indiscriminately at the destruction of entire cities or of extensive areas along with their populations is a crime against God and against man himself. It merits unequivocal and unhesitating condemnation.[6]

In an earlier letter[7] the American Bishops argued that even threats against enemy population centers were immoral, since they required a conditional intention to do what was immoral. But the moral principle on which they relied, that intention to do what is immoral is itself immoral, seems to me not to apply to the kind of self-frustrating intention which characterizes deterrent threats. For an intention to do something immoral is objectionable just because it is the last step the agent takes on the way to performance of an immoral action. The intention to retaliate which characterizes the nuclear deterrent strategy, however, is conditioned on an event (a Soviet attack) which it is believed, the very adoption of the intention makes unlikely. Thus, far from being anyone's last step on the way to performance of the act of retaliation, it is designed, among other things, precisely to make the act of retaliation unnecessary. In any case the point is moot, for several recent statements of official U.S. policy have explicitly stated that "for moral and political reasons, the United States does not target the Soviet civilian population as such."[8]

Second, even with the most scrupulous attention to the moral principle of non-combatant immunity, retaliatory threats raise both self-interested and moral worries. The very existence of the weapons raises the possibility, however remote, of accidents. Further, if deterrence should fail, there is always some danger that war would escalate to nuclear war and nuclear war to massive nuclear exchanges. And despite the technological advances which increase the accuracy of nuclear delivery vehicles, the co-location of military targets and civilian populations keeps alive the prospect of widespread collateral damage.

Deterrence: Inadequate for Long-term Peace

Nuclear deterrence, whether implemented through reliance on a policy of mutually assured destruction or on a policy of flexible response, may be the best of the currently available options, but it is just what the Bishops declare it to be, inadequate as a long-term basis for peace.[9] In saying this, they were echoing sentiments which had been expressed by Pope John Paul II the year before. In a message to the U.N. he had said:

> In current conditions, "deterrence" based on balance, certainly not as an end in itself, but as a step on a way toward progressive disarmament, may still be judged morally acceptable. Nonetheless, in order to assure peace, it is indispensable not to be satisfied with this minimum which is always susceptible to the real danger of explosion.[10]

The dissatisfaction with nuclear deterrence is not, of course, confined to the clergy. In a recent conversation, President Reagan expressed his dissatisfaction with the present state of affairs:

> Think of it. You're sitting at that desk. The word comes that they (the missiles) are on their way. And you sit here knowing that there is no way, at present, of stopping them. So they're going to blow up how much of this country we can only guess at, and your response can be to push the button before they get here so that even though you're all going to die, they're going to die too. . . .There's something so immoral about it.[11]

The Bishops urge that we "move. . .in a new direction, toward a national policy and an international system which more adequately reflect the values and vision of the kingdom of God."[12] The President put the same point as follows: "The human spirit must be capable of rising above dealing with other nations and human beings by threatening their existence."[13]

The Strategic Defense Policy

These remarks lead us into the third policy option which might be adopted in response to the nuclear dilemma—strategic defense. In the opening paragraph of my essay, I argued that the dilemma we face was created by the coupling of an invulnerable delivery vehicle—the ballistic missile—and a devastating warhead. A policy of strategic defense would be designed to make the delivery vehicle vulnerable.

Such a policy would not be a panacea. It would not solve the problems of poverty or racism. It would not even *solve* the problem of national security, but it would nevertheless solve one particularly intractable aspect of that last problem in a way preferable to the current policy for reasons both of morality and of self-interest.

The advantages of a defensive over a retaliatory policy seem clear. Defense against nuclear weapons raises no problems of intending (in any sense) what it would be immoral to do. Defense against nuclear weapons raises no problems of what to do if deterrence should fail. Defense against nuclear weapons does not leave us vulnerable to accidents. And perhaps most importantly, the ability to defend oneself against nuclear weapons opens the road to nuclear disarmament, by removing the instability of small arsenals and by removing the incentive to cheat on disarmament agreements. These are, of course, precisely the kinds of advantages cited by President Reagan in proposing that we look into strategic defense. At one point he asked, "Would it not be better to save lives than to avenge them?" and at another he points out that strategic defense could "pave the way for arms control measures to eliminate the weapons themselves."

SDI: Morally Superior Over the Alternatives

Does strategic defense have disadvantages? Two are commonly cited.

First, critics object that it undermines the only really successful nuclear arms control treaty we have ever achieved. It is true that, though research into the possibility of strategic defense would not violate any provision of SALT I, deployment of such a system, at least on the scale being discussed here, would do so. But it is not clear that it is for that reason bad. Arms control treaties are, after all, not ends in themselves; they are means to preserving national security and international peace, as they themselves make clear.[14] They should remain in force only so long as they help to achieve the end for which they were established. But SALT I makes sense only in the context of a strategy of mutually assured destruction, of vulnerable populations and invulnerable retaliatory forces. And that, as we have seen, is a policy which Bishops, Pope, and President all agree is acceptable only as an interim strategy!

Second, critics object that implementation of strategic defense would open a dangerous window of instability, a period in which the Soviet Union would see itself as having one last shot at the American nuclear arsenal before it became forever out of reach, protected by the strategic defensive shield. This is, of course, a danger. Careful co-ordination with our adversaries would be necessary to assure that we, and they, deploy strategic defensive systems in such a way that no such window of instability is opened. And the Reagan administration is explicit about the fact that the Strategic Defense Initiative should be "a cooperative effort with the Soviet Union, hopefully leading to an agreed transition toward effective non-nuclear defenses."[15] But the problem, though real, does not seem insuperable.

The real question seems to be, not whether strategic defense would be morally superior to deterrence by threat of nuclear retaliation, but whether it would work, and whether we can afford it. Both questions, however, seem premature. Strategic defense is not, at present, a possibility. President Reagan's Strategic Defense Initiative proposes only that we initiate "a long-range research and development program," to see just what the possibilities are. The research will itself, of course, not be free, but as President Reagan asks, "is it not worth every investment necessary to free the world from the threat of nuclear war?" To which we might add, given what we have said about the moral superiority of this strategy over the alternatives, would it be morally permissible to ignore what may be the only road to the elimination of nuclear weapons?

---

[1] *Gaudium et Spes,* paragraph 77.

[2] *The Challenge of Peace: God's Promise and Our Response,* para. 174.

[3] Speech on Defense Spending and Defensive Technology, March 23, 1983.

[4] *Ibid.,* para. 175. Emphasis in original.

[5] *Ibid.,* para 189. Indeed their failure to recognize that their emphasis on the good of disarmament and their concern for avoiding instability run in somewhat different directions is one of the unresolved tensions of the pastoral.

[6] *Gaudium et Spes,* para 80.

[7] *To Live in Christ Jesus* (1976).

[8] Letter to Mr. William Clark, then national security advisor, to Cardinal Bernardin, January 15, 1983. See also, Secretary of Defense Casper Weinberger's *Annual Report to the Congress,* February 1, 1983, p. 55.

[9] *Op. Cit.,* para. 186.

[10] Message, U.N. Special Session, 1982.

[11] Hugh Sidey, "The Presidency," *Time,* January 28, 1985, p. 29.

[12] *Op. Cit.,* para. 134.

[13] Speech of March 23, 1983.

[14] Cf. "Basic Principles of Relations between the United States and the Union of Soviet Socialist Republics," *Department of State Bulletin,* June 26, 1972, pp. 898-899, signed by President Nixon and General Secretary Brezhnev at the 1972 Moscow Summit which accompanied signature of the SALT I accord. This document states in part that "The USA and the USSR regard as the ultimate objective of their efforts the achievement of general and complete disarmament and the establishment of an effective system of international security."

[15] Mr. Paul Nitze, "On the Road to a More Stable Peace," A Speech before the Philadelphia World Affairs Council, February 20, 1985.

STAR WARS
AND MORAL CHOICES

THE MORAL CASE
AGAINST THE
STRATEGIC DEFENSE INITIATIVE

George E. Griener

George E. Griener, S.J., wrote this article in his capacity as a theology teacher at Loyola University in New Orleans, Louisiana. His article appeared in America, *a weekly, Catholic Jesuit publication on moral and public affairs.*

Points to Consider:

1. How was SDI originally portrayed? How did this vision change?
2. Does the SDI program enhance or weaken security? Provide evidence to support your answer.
3. Is SDI cost/benefit effective? Why or why not?
4. Why does the author conclude that SDI is not morally superior?

Given the data at our disposal, we can only conclude that the Strategic Defense Initiative does not present us with a morally superior alternative to the dilemma of nuclear deterrence.

The Strategic Defense Initiative (SDI) or "Star Wars" remains a sore point between the United States and the Soviet Union: The ABM Treaty (Anti-Ballistic Missile) of 1972 prohibits certain testing essential for SDI development. Moral momentum must be maintained in order to assure that the disarmament process does not run out of steam.

That moral momentum seems to stem from the early part of this decade. Public debate of our defense policy had assumed strong and politically important ethical dimensions by 1982. From the development of the neutron bomb to the installation of Pershing II and cruise missiles in western Europe, to the basing mode of MX missiles in the United States, hardly a single issue escaped political and moral analysis— sometimes naive and cliched, but more often penetrating and incisive. Open discussion surrounding the formulation of the U.S. Catholic bishops' pastoral letter, "The Challenge of Peace: God's Promise and Our Response" (1983), helped focus that debate.

The bishops, following the lead of Pope John Paul II, and in unison with their fellow bishops in France and West Germany, granted deterrence conditional acceptability, but only as long as it was one element of a larger strategy of progressive nuclear disarmament. Pastoral statements focusing on nuclear deterrence were issued by other church groups in the United States, Canada, England, and countries on the European continent.

The Reagan Administration was alarmed at the critical moral stance of opposition groups in the United States vis-a-vis nuclear deterrence and at the political consequences it might have. There is evidence that this public discussion and the anti-nuclear demonstrations played an important role in President Reagan's decision, just six weeks before the final publication of the U.S. bishops' pastoral, to proclaim his vision of a morally superior alternative to the strategy of nuclear deterrence: the Strategic Defense Initiative, perhaps best known as "Star Wars."

The Star Wars Vision

President Reagan's speech on March 23, 1983, was surprising not only in the content of its vision of an umbrella of defense, of a technological shield intended to render nuclear weapons "impotent and obsolete." It was surprising also in its implied acceptance of the moral position of the nuclear weapons opponents: namely, that the strategy of nuclear deterrence was inhumane and morally unacceptable. "I have become more and more deeply convinced," said President Reagan, "that the human spirit must be capable of rising above dealing with other nations and human beings by threatening their existence. . . .Would it not be better to save lives than to avenge them?. . .I call upon the scientific community who gave us nuclear weapons to turn their great talents to the cause of mankind and world peace; to give us the means of rendering these nuclear weapons impotent and obsolete."

SDI was portrayed as a multilayered system of defense that would intercept and destroy all missiles and warheads aimed at the United States. The U.S. mainland, population centers as well as military installations, would be protected from an enemy attack. The perceived futility of attacking the United States by using nuclear weapons carried by ballistic missiles would itself deter any aggressor from launching an attack. By implication, there would be no need for the United States to threaten use of nuclear weapons in response. Defense would replace deterrence as our security strategy; SDI would be strategically, politically, and morally superior to its predecessor.

Moreover, to forestall criticism that such a shield would give the United States an unfair strategic advantage, President Reagan vowed to share space-defense research data with the Russians. This would maintain relative technical parity between the two nations and, presumably, avoid destabilizing the precarious balance of offensive and defensive power the two possess.

From an ethical perspective, this vision of SDI seemed to be an attractive alternative to the strategy of nuclear deterrence. Peace will never be achieved by technical means alone; and SDI would not address

Illustration by John Houser. Reprinted with permission of *The People*.

the problems of the regional conflicts that aggravate superpower relations and lead to unjustified bloodshed and human misery—Central America, the Middle East, and Afghanistan are but three examples. But SDI, fully deployed, promised to reduce the threat of an all-out nuclear exchange between the Soviet Union and the United States. If, by some snap of the fingers, the entire scenario could suddenly be realized and set in place, then most would applaud Star Wars as having achieved a moral breakthrough.

An Altered Vision

Since 1983 this vision of SDI has undergone critical analysis, some profound changes and substantial redirection. Few scientists are willing to accept the model of a continental umbrella capable of rendering all nuclear weapons "impotent and obsolete." Independent scientific studies as well as Congressional reports underscore serious technical problems that need far more research before an evaluation can be made about the potential viability of many components of the system. The American Physical Society report, issued in April 1987, is perhaps the most recent to point to "serious gaps in the scientific and engineering understanding of many issues associated with the development of these technologies."

Even after enormous political pressure from the United States, only a few Western nations have acceded to the U.S. invitation to join the SDI research project. Those who agreed placed conditions on their cooperation: The terms of the 1972 ABM Treaty must be respected. Of course there is considerable debate as to whether components of SDI can be tested without violating that ABM agreement. Recognizing this, the Reagan Administration has pushed for a "broad" interpretation of the 1972 treaty, an interpretation that even many of the treaty's original U.S. negotiators find erroneous. Not only the Soviet Union but our Western allies and the U.S. Congress itself are pressing the President to remain within the constraints of the so-called strict interpretation of the treaty. The recently passed 1988 Defense Appropriations Bill prohibits the Administration from using the broad interpretation throughout the coming fiscal year, thus putting off a final showdown over treaty interpretation for at least another year.

The President's intention to share space-defense technical information with the Soviet Union is not likely to be fulfilled. Current government policy regarding publication of space-defense-related discoveries is so restrictive that research industries in allied countries have hesitated to bid for contracts with Star Wars: Their inability to make use of research data in domestic and nonmilitary applications would render their participation in Star Wars research financially unprofitable. Scientists have complained that the secrecy surrounding many experiments, which are not in themselves exclusively "military" in importance, will jeopardize the exchange of ideas necessary for the progress of basic scientific research.

Military strategists are now speaking about a much more modest anti-missile defense than the one President Reagan outlined. The optimists among them, who still foresee a nationwide umbrella, are predicting only a 90 percent to 95 percent "kill" reliability, others no more than 50 percent. But if 50 percent, or even 10 percent of the warheads directed against the United States slip through the shield, that would be enough to destroy most major cities, scores of millions of their inhabitants and effectively cripple the infrastructure of American society. Moreover, the long coastlines of the United States make it especially vulnerable to attack from submarine launched ballistic or cruise missiles, against which SDI offers little or no defense.

This high probability of missile "leakage" would require that the United States maintain a sizable nuclear missile strike force to provide supplemental deterrent credibility. Thus SDI would complement but not replace our reliance on nuclear weapons for keeping "the peace." Nuclear deterrent—namely, the intent and the threat to use our nuclear weapons in retaliation—would remain the essential element of our security strategy.

These changes alter the configuration of the original proposal and also the initial moral evaluation of the program. They demand a moral reappraisal of SDI.

An Ethical Appraisal

The complexities of the nuclear standoff between the United States and the Soviet Union make a moral assessment of SDI a very complex undertaking. SDI has to be seen within the larger context of our nation's defense strategy, foreign policy, and the goal of further disarmament. We can only expect a prudential moral judgment based on a sober evaluation of the parameters that affect the situation, evaluations about which experts disagree: the strategic, defensive, political, and economic consequences of the program.

We must also know precisely what it is that we are trying to evaluate. It might be argued that SDI has been responsible for making the Soviet Union more flexible in the disarmament talks and has already demonstrated its worth as a "bargaining chip." But it is also clear that President Reagan envisions SDI as more than that, and intends to deploy it as soon as its components should become operative.

Several questions are in order: Does SDI elevate or relax world tensions? Does it decrease or increase the likelihood of a nuclear exchange between the Soviet Union and the United States? Does it simplify or complicate the discussion of arms control and disarmament? Is it cost-effective, or will it unduly militarize the ever more limited financial resources of the United States? Does it represent a radical shift from a strategy of nuclear deterrence to a truly defensive application of technology? Or is it the first phase of a third generation of offensive nuclear weaponry?

Does the SDI Program Enhance or Weaken Security?

What impact would even a partially deployed defensive system have on the precarious balance of nuclear forces between the superpowers, a balance that is at least partially responsible for the prevention of a nuclear attack by one of the parties on the other?

Soviet opposition to SDI does not automatically provide the United States with a moral justification for the program. Even an enemy has legitimate concerns about his security. Experts recognize the technological superiority of the United States in the area of microelectronics and computer hardware, not to mention in many specific fields relating to the technology of space defense. We can thus presume that the Soviet Union has legitimate concerns about its own security in the face of the massive U.S. undertaking to provide for itself even a limited defensive shield against nuclear missiles.

In their 1983 pastoral, the U.S. bishops pointed out that "preserving stability requires a willingness by both sides to refrain from deploying weapons that appear to have a first-strike capability" (no. 163). Given a parity between the superpowers in the number of warheads, the side with the defensive advantage would have an effective superiority. Even if SDI itself is characterized as a defensive weapon system, it could be seen as enhancing the first-strike capabilities of the West when integrated into a larger nuclear deterrent complex.

The bishops also argued that "each proposed addition to our strategic system or change in strategic doctrine must be assessed precisely in light of whether it will render steps toward 'progressive disarmament' more or less likely" (no. 188.3). In preparation for the December summit, the Russians moderated their voiced objections to SDI, but there is reason to believe that this was a tactical move. Mikhail Gorbachev sees no point in pressing an issue that can be addressed later or that U.S. budgetary restraints may well eliminate as a major problem in the future. But in late December, national security adviser Lieut. Gen. Colin Powell admitted that U.S. testing of SDI components would dash hope for further disarmament and might precipitate Soviet increases of its strategic defense forces. SDI would accelerate the Soviet Union's drive to accelerate research on a similar defense system and to explore all possible ways of countering SDI through technological means.

Budget: Is It Cost/Benefit Effective?

U.S. arms negotiator Paul Nitze and British Foreign Secretary Sir Geoffrey Howe have asked whether SDI would be able to provide defense against nuclear weapons more cheaply than the Soviet Union could produce warheads or decoys to overwhelm the defense system. The system would be very vulnerable. Damaging the space-based components of SDI would be relatively simple and inexpensive. And there are innumerable methods of jamming or overwhelming the system with foils and decoys. Finally, the Soviet Union is developing rockets with shorter boost-phase times, making more complicated the planned first level of SDI—attacking hostile rockets before the multiple warheads are released on their separate paths.

Cost estimates of a fully deployed SDI system of defense range from several hundred billion dollars upwards to $1 trillion. If a less-than-perfect SDI shield will require maintaining modern deterrent nuclear forces as well, then there is even more reason not to accelerate the research project of SDI. Considering the minimal benefits that seem to be achievable at present, and indeed the risks, strategic and political, involved in going ahead with the project, it would be irresponsible to earmark a significant percentage of the budget for SDI.

The security of the United States depends not only on sophisticated weapons systems, but on the health of its citizens and an educated electorate, as well as on the equitable distribution of our resources and a just economic world order. The stock market problems of last fall have underlined the need to reduce substantially our nation's deficit. In the face of budgetary restraints it will increasingly become a moral imperative to designate long-range national priorities justly and prudently. From a cost/benefit perspective there is no moral justification for pursuing the costly deployment of SDI at this time.

Arms Race: New Generation of Nuclear Weapons?

In analyzing the morality of nuclear deterrence, the United States, French, and West German bishops stressed that the deterrent policy was "conditionally acceptable" only as long as there was no attempt to achieve "superiority" over the opponent. The French bishops put it this way: "Deterrence is attained from the moment when the formulated threat makes aggression by a third party unreasonable." Deterrence is no longer morally tolerable beyond that point.

Recently released Congressional testimony suggests that policy planners see the SDI program not just as a defensive measure, but as ushering in a third generation of nuclear weapons. Proponents argue that these "nuclear directed-energy" weapons are more precise and more accurate because the destructive force is not unleashed in an expanding spherical configuration, but aimed and directed at the intended target. Prof. Edward Teller's Excalibur Project would be an example.

If military strategists are considering offensive use of the weapons being developed in the SDI program, then we are facing a new phase of the arms race. Not only would SDI fail to replace existing strategic deterrent forces with defensive technology, but the new weaponry would be understood as possessing potential offensive application as well.

Conclusion: SDI Not Morally Superior

A sober evaluation of SDI has led to serious reservations about its touted "moral superiority." 1) Most analysts and scientists see no hope for a perfect shield against nuclear weapons. At best, SDI would complement but not replace nuclear deterrence as the foundation of our security. 2) There is no persuasive evidence that SDI would significantly reduce the threat to the U.S. mainland and population of a large-scale nuclear attack. 3) Moreover, the enormous costs of implementing the system would jeopardize the funding of programs equally important in maintaining our welfare and security: education, health care, and overcoming the political and economic inequities in the world order. 4) Finally, attempts to direct SDI research toward offensive weapon application pose the danger of accelerating the arms race, which we have vowed to avoid. Given the data at our disposal, we can only conclude that the Strategic Defense Initiative does not present us with a morally superior alternative to the dilemma of nuclear deterrence.

SDI was proposed at a time not only when objections were being raised about the morality of nuclear deterrence, but when there seems to have been little interest on the part of the Reagan Administration to pursue diplomatic resolutions of many of the world's problems. Little wonder that a technological solution looked so attractive. Perhaps a new realism and the positive experience of hard diplomatic bargaining will restore confidence in a broader, more integral solution to the danger of war and advance the goal of nuclear disarmament and world peace. One should dream dreams of peace that are down to earth.

WHAT IS EDITORIAL BIAS?

This activity may be used as an individualized study guide for students in libraries and resource centers or as a discussion catalyst in small group and classroom discussions.

The capacity to recognize an author's point of view is an essential reading skill. The skill to read with insight and understanding involves the ability to detect different kinds of opinions or bias. Sex bias, race bias, ethnocentric bias, political bias, and religious bias are five basic kinds of opinions expressed in editorials and all literature that attempts to persuade. They are briefly defined below.

Five Kinds of Editorial Opinion or Bias

SEX BIAS—The expression of dislike for and/or feeling of superiority over the opposite sex or a particular sexual minority

RACE BIAS—The expression of dislike for and/or feeling of superiority over a racial group

ETHNOCENTRIC BIAS—The expression of a belief that one's own group, race, religion, culture, or nation is superior. Ethnocentric persons judge others by their own standards and values

POLITICAL BIAS—The expression of political opinions and attitudes about domestic or foreign affairs

RELIGIOUS BIAS—The expression of a religious belief or attitude

---

Guidelines

1. From the readings in Chapter Four, locate five sentences that provide examples of editorial opinion or bias.

2. Write down each of the above sentences and determine what kind of bias each sentence represents. Is it *sex bias, race bias, ethnocentric bias, political bias, or religious bias?*

3. Make up one sentence statements that would be an example of each of the following: *sex bias, race bias, ethnocentric bias, political bias, and religious bias.*

4. See if you can locate five sentences that are factual statements from the readings in Chapter Four.

APPENDIX

Strategic Defense Initiative Glossary

(A guide to technical terms for use as a reference in understanding articles on and discussions of the SDI.)

ABM System Antiballistic missile, a system to counter strategic ballistic missiles or their elements during flight.

Acquisition Detection of a potential target by the sensors of a weapons system.

Active Sensor One that illuminates a target, producing return secondary radiation, which is then detected in order to track and/or identify the target.

Anti-Satellite Weapon (ASAT) A weapon to destroy satellites in space.

Anti-Simulation Deceiving adversary sensors by making a strategic target look like a decoy.

Area Defense An ABM defense covering a large area. Usually implies the capability to protect "soft" (i.e., not hardened missile silos or bunkers) targets.

Ballistic Missile Any missile that does not rely upon aerodynamic surfaces to produce lift and consequently follows a ballistic path when thrust is terminated. Ballistic missiles typically operate outside the atmosphere for a substantial portion of their flight path and are unpowered during most of their flight.

Ballistic Missile Defense (BMD) A defense system that is designed to protect territory from attacking ballistic missiles. Usually conceived as having several independent layers.

Booster The rocket that "boosts" the payload to accelerate it from the earth's surface into a ballistic path, during which no additional force is applied to the payload.

Boost Phase The portion of a missile flight during which the payload is accelerated by the large rocket motors. For a multiple-stage rocket, the boost phase involves all motor stages. For ICBMs, this phase typically lasts from 3 to 5 minutes, but studies indicate that reductions to the order of 1 minute could be possible.

Brightness In the SDI context, the amount of power that can be delivered per unit solid angle by a directed-energy weapon.

Bus Deployment Phase The portion of a missile flight during which multiple warheads are deployed on different paths to different targets (also referred to as the post-boost phase). The warheads on a single missile are carried on a platform, or "bus" (also referred to as a post-boost vehicle), which has small rocket motors to move the bus slightly from its original path.

Command Guidance The steering and control of a missile by transmitting commands to it.

Counter-Countermeasures Measures taken by the defense to defeat offensive countermeasures.

Counterforce The employment of strategic nuclear forces in an effort to destroy, or disable, selected military capabilities of an enemy force.

Countermeasures In the SDI context, measures taken by the offense to overcome aspects of the BMD system.

Cruise Missile A missile traveling within the atmosphere at aircraft speeds and, usually, low altitude, whose path is preprogrammed. It is capable of achieving high accuracy in striking a distant target. It is maneuverable during flight, is constantly propelled, and therefore does not follow a ballistic path. Cruise missiles may be nuclear armed.

Dazzling In the SDI context, the temporary blinding of a sensor by overloading it with an intense signal of electromagnetic radiation, e.g., from a laser or a nuclear explosion.

Decoy A device that is constructed to look and behave like a nuclear-weapon-carrying warhead, but which is far less costly, must less massive, and can be deployed in large numbers to complicate defenses.

Defensive Satellite Weapon (DSAT) A device that is intended to defend satellites in space by destroying attacking ASAT weapons.

Digital Processing The most familiar type of computing, in which problems are solved through the mathematical manipulation of streams of numbers.

Directed Energy Energy in the form of particle or laser beams that can be sent long distances at nearly the speed of light.

Directed-Energy Weapon A weapon that kills its target by delivering energy to it at or near the speed of light. Includes lasers and particle beam weapons.

Early Warning In the SDI context, the early detection of an enemy ballistic missile launch, usually by means of surveillance satellites and long-range radar.

Electromagnetic Radiation A form of propagated energy, arising from electric charges in motion, that produces a simultaneous wavelike variation of electric and magnetic fields in space. The highest frequencies (or shortest wavelengths) of such radiation are possessed by gamma rays, which originate from processes within atomic nuclei. As one goes to lower frequencies, the electromagnetic spectrum includes X-rays, ultraviolet light, visible light, infrared light, microwaves, and radio waves.

Endoatmospheric Taking place within the earth's atmosphere, generally considered as occurring at altitudes below 100 kilometers.

Ephemeris A collection of data about the predicted positions (or apparent positions) of celestial objects, including artificial satellites, at various times in the future. A satellite ephemeris might contain the orbital elements of satellites and predicted changes in these.

Exoatmospheric Taking place outside the earth's atmosphere, generally considered as occurring at altitudes above 100 kilometers.

Fast-Burn Booster A ballistic missile that can burn out much more quickly than current versions, possibly before exiting the atmosphere entirely. Such rapid burnout complicates a boost-phase defense.

Fission The breaking apart of the nucleus of an atom, usually by means of a neutron. For very heavy elements, such as uranium, a significant amount of energy is produced by this process. When controlled, this process yields energy which may be extracted for civilian uses, such as commercial electric generation. When uncontrolled, energy is liberated very rapidly: such fission is the energy source of uranium- and plutonium-based nuclear weapons; it also provides the trigger for fusion weapons.

Functional Kill The destruction of a target by disabling vital components in a way not immediately detectable, but nevertheless able to prevent the target from functioning properly. An example is the destruction of electronics in a guidance system by a neutral particle beam.

Fusion More specifically, nuclear fusion. The fusing of two atomic nuclei, usually of light elements, such as hydrogen. For light elements, energy is liberated by this process. Hydrogen bombs produce most of their energy by the fusion of hydrogen into helium.

Gamma-Ray Laser A laser which generates a beam of gamma rays; also called a graser. A gamma-ray laser, if developed, would be a type of X-ray laser; although it would employ nuclear reactions, it need not (but might) employ nuclear fission or fusion reactions or explosions.

Gamma Rays X-rays emitted by the nuclei of atoms.

Hard Kill Destruction of a target in such a way as to produce unambiguous visible evidence of its neutralization.

Homing Device A device, mounted on a missile, that uses sensors to detect the position or to help predict the future position of a target, and then directs the missile to intercept the target. It usually updates frequently during the flight of the missile.

Hypervelocity Gun A gun that can accelerate projectiles to 5 kilometers per second or more, for example, an electromagnetic, or rail, gun.

Infrared Sensor A sensor to detect the infrared radiation from a cold body such as a missile reentry vehicle.

Intercept The act of destroying a target.

Ionization The removal or addition of one or more electrons to a neutral atom, forming a charged ion.

Isotropic Independent of direction; referring to the radiation of energy, it means "with equal intensity in all directions," i.e., omnidirectional.

Isotropic Nuclear Weapon (INW) A nuclear explosive which radiates X-rays and other forms of radiation with approximately equal intensity in all directions. The term "isotropic" weapon is used to distinguish it from a nuclear directed-energy weapon.

Joule A Systeme Internationale unit of energy. One kilowatt-hour is 3.6 million joules.

Kinetic Energy The energy from the momentum of an object.

Kinetic-Energy Weapon A weapon that uses kinetic energy, or energy of motion, to kill an object. Weapons that use kinetic energy are a rock, a bullet, a nonexplosively armed rocket, and an electromagnetic railgun.

Laser A device for generating coherent visible or infrared light.

Laser Designator The use of a low-power laser to illuminate a target so that a weapon equipped with a special tracker can home in on the designated target.

Laser Imaging A new technology where a laser beam can be used in a way similar to a radar beam to produce a high-quality image of an object.

Laser Tracker The process of using a laser to illuminate a target so that specialized sensors can detect the reflected laser light and track the target.

Layered Defenses The use of several layers of BMD at different phases of the missile path. Each layer is designed to be as independent as possible of the others, and each would probably use its own, distinctive set of missile defense technologies.

Leakage The percentage of warheads that get through a defensive system intact and operational.

Megawatt One million watts; a unit of power. A typical commercial electric plant generates about 500 to 1,000 megawatts.

Midcourse Phase The long period of a warhead's flight to its target after it has been dispensed from the post-boost vehicle until it reenters the atmosphere over its target. This phase lasts up to 20 minutes.

Miniature Homing Vehicle (MHV)/Miniature Vehicle(MV) An air-launched direct-ascent ("pop-up") kinetic-energy ASAT weapon currently being developed and tested by the U.S. Air Force.

Multiple Independently-Targetable Reentry Vehicle (MIRV) One of several reentry vehicles on the same post-boost phase vehicle that can be independently placed on a ballistic course towards a target after completion of the boost phase.

Multiple Reentry Vehicle The reentry vehicle of a ballistic missile which is equipped with multiple warheads but which does not have the capability of independently directing the reentry vehicles to separate targets.

Neutral-Particle Beam An energetic beam of neutral atoms (no net electric charge). A particle accelerator moves the particle to nearly the speed of light.

Obscurant A material (e.g., smoke or chaff) used to conceal an object from observation by a radio or optical sensor. Smoke may be used to conceal an object from observation by an optical sensor, and chaff may be used to conceal an object from observation by a radio sensor (e.g., radar).

Particle Beam A stream of atoms or subatomic particles (electrons, protons, or neutrons) accelerated to nearly the speed of light.

Passive Sensor A sensor that only detects radiation naturally emitted (infrared radiation) or reflected (sunlight) from a target.

Payload The weapons and penetration aids carried by a delivery vehicle.

Post-Boost Phase The phase of a missile path, after the booster's stages have finished firing, in which the various reentry vehicles are independently placed on ballistic paths towards their targets. In addition, penetration aids are dispensed from the post-boost vehicle. The length of this phase is typically 3 to 5 minutes, but could be drastically reduced.

Post-Boost Vehicle The portion of a rocket payload that carries the multiple warheads and maneuvering capability to place each warhead on its final path to a target (also referred to as a "bus").

Radar A technique for detecting targets in the atmosphere or in space by transmitting radio waves (e.g., microwaves) and sensing the waves reflected by objects. The reflected waves (called "returns" or "echos") provide information on the distance to the target and the velocity of the target and may also provide information about the shape of the target. (Originally an acronym for "Radio Detection and Ranging.")

Radiant Energy The energy from radiation such as electrons, protons, or alpha particles.

Reaction Decoy A decoy deployed only upon warning or suspicion of imminent attack.

Redout The blinding or dazzling of infrared detectors due to high levels of infrared radiation produced in the upper atmosphere by a nuclear explosion.

Reentry The return of objects, originally launched from Earth, into the atmosphere.

Reentry Vehicle (RV) In the SDI context, reentry vehicles are small containers containing nuclear warheads. They are released from the last state of a booster rocket or from a post-boost vehicle early in the ballistic path. They are thermally insulated to survive rapid heating during the high velocities of reentry into the atmosphere, and are designed to protect their contents until detonation at their targets.

Robust In the SDI context, indicating the ability of a system to endure and perform its mission against a reactive adversary. Also used to indicate ability to survive under direct attack.

Sensors Electronic instruments that can detect radiation from objects at great distances. The information can be used for tracking, aiming, discrimination, attacking, kill assessment, or all of the above. Sensors may detect any type of electromagnetic radiation or several types of nuclear particles.

Simulation The art of making a decoy look like a more valuable strategic target (cf. anti-simulation).

Smoke An obscurant which may be used in the atmosphere or in space to conceal an object from observation by an optical sensor.

Space Mines Hypothetical devices that can track and follow a target in orbit, with the capability of exploding on command or by pre-program, in order to destroy the target.

SS-18 Largest ICBM in current Soviet inventory, credited with carrying 10 reentry vehicles, but capable of holding many more.

Terminal Phase The final phase of a ballistic missile path, during which warheads and penetration aids reenter the atmosphere.

Threat The anticipated inventory of enemy weapons. In the SDI context, the inventory is of nuclear weapons and their delivery systems, as well as of decoys, penetration aids, and other BMD countermeasures.

Threat Clouds Dense concentrations of both threatening and non-threatening objects. The defense must distinguish between them.

Tracking The monitoring of the course of a moving target. Ballistic objects may have their tracks predicted by the defensive system, using several observations and physical laws.

Transition In the SDI context, the period in which the world strategic balance would shift from offense-dominance to defense-dominance.

Warhead The part of a missile, projectile, torpedo, rocket, or other munition containing either the nuclear or the thermonuclear system, high explosive system, chemical or biological agents, or inert materials intended to inflict damage.

X-ray Laser A laser which generates a beam or beams of X-rays. Also called an "x-raser" or "XRL."

X-rays Electromagnetic radiation having wavelengths shorter than 10 nanometers (10 billionths of a meter).

BIBLIOGRAPHY (ANNOTATED)

This annotated bibliography is arranged in five topics in the following order: the status of the U.S. program, the Soviet program, arms control and the SDI, the strategic implications of SDI, and allied views.

The Status of the U.S. Program

1. Broad, William. "Reagan's Star Wars Bid: Many Ideas Converging." *New York Times,* March 4, 1985.

   The role of Dr. Edward Teller, and third-generation weapons, in President Reagan's "Star Wars" proposal. The Bethe-Teller dispute as it parallels the proponents of political viz technological solutions to the world's ills. How Reagan's "kitchen cabinet" split over the tough questions concerning SDI. Questions whether the goal is defense or nuclear superiority.

2. Carter, Ashton B. "Directed Energy Missile Defense in Space." Background Paper. Prepared for the Office of Technology Assessment, Congress of the United States, April 1984.

   A description and assessment of current concepts in directed-energy ballistic missile defense, prepared for members of Congress and the general public. Also discusses the arms control implications of such a system.

3. Fletcher, James C. "The Technologies for Ballistic Missile Defense." *Issues in Science and Technology,* Fall 1984, 15-29.

   Chairman of the Fletcher Commission posits his views on the technology of ballistic missile defense (BMD) with recommendations for a vigorous R&D program. Defines "missing" technologies. Describes technological challenges as "great but not insurmountable." Sees SDI as an opportunity for strengthening deterrence.

4. Gerry, Edward T. "The Strategic Defense Initiative."

   Paper presented at the Conference on Nuclear Deterrence—New Risk, New Opportunities, University of Maryland, College Park, Maryland, September 5, 1984.

   Provides explanation for SDI as primary basis for deterrence by removing the first strike capability of the other side. Explores technological advances in "layered" defense.

5. Gerry, Edward T. "The Strategic Defense Initiative." Paper presented to the American Physical Society, Boston, Massachusetts, November 1, 1984.

Answers the question: "Why change U.S. nuclear strategy from MAD?" with a variety of answers, relating to SDI and ultimate stability. Also addresses the technology issues.

6. Golden, Frederic. "Special Report: Star Wars." *Discover,* September 1985, 28-42.

A status report on research on the SDI.

7. Sagan, Carl. "The Case Against SDI." *Discover,* September 1985, 66-75.

Countering Edward Teller, this famed author and astronomer argues against SDI in terms of lack of feasibility and humanitarian concerns.

The Soviet Program

1. Mohr, Charles. "What Moscow Might Do in Replying to 'Star Wars.'" *New York Times,* March 6, 1985, p. 1.

Starts with the contention that Soviets and U.S. are equals in space- and land-based BMD. The Soviet program from the perspective of Sovietologists in the U.S. Some analysts fear Soviets closer to nation-wide BMD than the U.S.

2. Nitze, Paul H. "SDI: The Soviet Program." Speech to the Chautauqua Conference on Soviet-American Relations, Chautauqua, New York, June 28, 1985. Current Policy no. 717. Department of State, Bureau of Public Affairs.

An analysis of the Soviet version of SDI and other ABM-related activities by the Special Advisor to the President and the Secretary of State on Arms Control Matters. A technological discussion of lasers, particle beams, radio frequency, kinetic energy, and other weapons being produced by the Soviet Union.

3. U.S. Department of Defense. *Soviet Military Power.* Washington, DC: U.S. Government Printing Office, 1985.

Describes Soviet concept of "layered defense" and its individual components. A highly definitive status report for the Soviet program, providing details of treaty violations. Also describes Soviet "passive defense" system.

4. U.S. Department of State, Bureau of Public Affairs. "Security and Arms Control: The Search for a More Stable Peace." September 1984.

A brief summary of U.S. and Soviet status in the area of missile defense and how it accords with arms control treaties.

Arms Control and the SDI

1. Adelman, Kenneth L. "SDI: Setting the Record Straight." Speech to the Baltimore Council on Foreign Affairs, August 7, 1985. Current Policy no. 730. Department of State, Bureau of Public Affairs.

   The Director of the U.S. Arms Control and Disarmament Agency lists "impossible" situations akin to SDI as a take-off for discussion of questions regarding this proposal, most particularly the congruence of SDI with the ABM Treaty, the Non-Proliferation Treaty, and arms control generally.

2. "Aerospace Experts Challenge ASAT Decision." *Science,* May 19, 1984, 693-696.

   Describes the ASAT "war" in Congress as well as the President's position. Investigates the issue of verification and how ASAT fits in with arms control.

3. Durch, William J., editor. *National Interests and the Military Use of Space.* Cambridge, MA: Ballinger Publishing Co., 1984.

   Takes a close look at the strategic significance and implications for security of space weapons.

4. Gelb, Leslie H. "Weapons in Space: The Controversy Over 'Star Wars'." *New York Times,* March 3, 1985, p. 1.

   Compares "myth" and "reality" of the proposed SDI. Calls it "the touchstone of loyalty to the President." Asks whether SDI will set off a major arms race. Touches on various officials, in and out of government, and their arguments for or against the system. Asks whether it encourages a first strike by the Soviets.

5. Meyer, Stephen M. "Soviet Military Programmes and the 'New High Ground'." *Survival,* September—October 1983, 204-215.

   Contends that the Soviets exceed U.S. in its space program and are accelerating their programs. Asks key questions regarding their militarization of space. Assesses impact of the Soviet space program on their global military involvement. Provides a review of early Soviet space weaponry. Posits overall conclusion on Soviet space program.

6. Nitze, Paul H. "SDI and the ABM Treaty." Commencement address before the Johns Hopkins School of Advanced International Studies, May 30, 1985. Department of State, Bureau of Public Affairs.

An explanation of SDI and its interaction with the ABM treaty and arms control generally, as well as an assessment of U.S. reappraisal towards strategic defense, by the Special Advisor to the President and the Security of State on Arms Control Matters.

7. Rivkin, David B., Jr. "What Does Moscow Think?" *Foreign Policy,* Summer 1985, 85-105.

Analyzes SDI from the Soviet perspective and sees a defense-oriented world as beneficial to the Soviet Union. Examines basic, historical Soviet strategic doctrine.

8. Shultz, George. "Arms Control: Objectives and Prospects." Speech to the Austin Council on Foreign Relations, Austin, Texas, March 28, 1985. Current Policy no. 675. Department of State, Bureau of Public Affairs.

An address by the Secretary of State on U.S. foreign policy, Soviet arms control violations, and U.S. objectives in Geneva in terms of long-term arms control goals. Also tackles SDI and the cost-effectiveness issue as well as the Soviet counterpart.

9. U.S. Congress, Office of Technology Assessment. *Arms Control in Space: Workshop Proceedings.* 1984.

Workshop focuses on anti-satellite (ASAT) weapons, exploring technical, diplomatic, military and policy aspects. It outlines points of general agreement among the experts, as well as areas of disagreement and topics for continuing research.

10. The White House. "The President's Strategic Defense Initiative." January 1985.

The President questions the basic assumptions of deterrence, in reproposing the SDI and speaks of the concomitant need for arms control. Speech contains explanation of the original SDI proposal by the President in March 23, 1983, speech. Presents concept of local defense to enhance deterrence, while dovetailing with arms control.

The Strategic Implications of SDI

1. Boffey, Philip M. "Dark Side of 'Star Wars': System Could Also Attack." *New York Times,* March 7, 1985, p. 1.

Critics allege that SDI could also have offensive capability, which would enable defensive capability to be held in reserve against retaliatory strike. Asks the question: What are the targets of SDI? Proponents claim SDI weapons are "precise weapons that destroy only precise targets."

2. Boffey, Philip M. "'Star Wars' and Mankind: Unforeseeable Directions." *New York Times,* March 8, 1985, p. 1.

   Captures some of the dialogue on Capitol Hill regarding the "Star Wars" program. Explores the status of research. Opponents see existence of nuclear weapons as guarantors of peace. Examines various "layers" of debate, both pro and con. Poses the problems of the transitional phase.

3. Drell, Sidney D., and Panofsky, Wolfgang K. H. "The Case Against: Technical and Strategic Realities." *Issues in Science and Technology,* Fall 1984.

   Explains the theory of mutual assured destruction (MAD) and the fact that "enhancement of stability" is of mutual interest as a backdrop to investigation of various "selling points" of SDI. Questions ultimate congruence of SDI and existing arms control treaties.

4. Hoffman, Fred. "The SDI in U.S. Nuclear Strategy." *International Security,* Summer 1985, 13-24.

   Claims that, while the U.S. has relied on MAD, the Soviets have swung the nuclear balance in their favor. Questions the Soviet belief in MAD and the "law of action and reaction." Calls for a prudent approach to SDI.

5. McNamara, Robert S., and Bethe, Hans A. "Reducing the Risk of Nuclear War." *Atlantic,* July 1985, 43-51.

   Authors question whether current policymakers believe President Reagan's adage that "a nuclear war cannot be won and must never be fought." They analyze what they term Star Wars I and II, posit restructuring of nuclear forces as an alternative, and comment on Nitze's Philadelphia Speech.

6. Nitze, Paul H. "On the Road to a More Stable Peace." Speech to the Philadelphia World Affairs Council, February 20, 1985. Current Policy no. 657. Department of State, Bureau of Public Affairs.

   The Special Advisor to the President and the Secretary of State on Arms Control Matters challenges basic tenets of deterrence and sets the stage for "non-nuclear defense that might make possible the eventual elimination of nuclear weapons." Discusses the cost-effectiveness issue and the problems inherent in the transitional period.

7. Schlesinger, James R. "Rhetoric and Realities in the Star Wars Debate." *International Security,* Summer 1985, 13-24.

   Criticizes proponents of SDI and their phrase, "the immorality of deterrence." Makes the claim that no defense will ever be impenetrable. Questions the concept of "mutual assured survival." Proposes that the best role for SDI is that of the classical bargaining chip.

8. Talbott, Strobe. Interview, March 1985. *Harvard International Review,* March/April 1985, 13-15.

   In arms control context, this *Time* magazine correspondent analyzes SDI and European views on the subject.

9. Teller, Edward. "The Case for the Strategic Defense Initiative." *Discover,* September 1985, 66-75.

   This renowned physicist argues the case for using defense as a deterrent in place of mutual assured destruction (MAD).

10. U.S. Congress, Office of Technology Assessment. *Anti-Satellite Weapons, Countermeasures, and Arms Control.* September 1985.

    Study done at request of House Armed Services Committee and Senate Foreign Relations Committee which assesses implications of ballistic missile defense technologies and others similar to it. Report examines options open to U.S. for countering Soviet military satellite capabilities and measures to limit ASAT threat. Also examines pros and cons of various arms control ideas for space weapons.

11. U.S. Department of State, Bureau of Public Affairs. "The Strategic Defense Initiative." Special Report no. 129. June 1985.

    An explanation of SDI in the strategic context, most specifically in terms of deterrence. It presents a view of Soviet status in the strategic balance and how that status effects deterrence. Answers twelve questions relating to SDI.

Allied Views

1. Alterman, Eric R. "European Leaders' Quandray." *New York Times,* July 11, 1985, p. 23.

   Expresses European fears of SDI in terms of their vulnerability, in the event of attack, and in the world commercial marketplace.

2. Drodziak, William. "Bonn Delegation Comes to U.S. to Discuss 'Star Wars' Participation." *Washington Post,* September 5, 1985.

   Highlights U.S. tour by German delegation, investigating the technological opportunities of SDI. Assesses impact of French "Eureka" concept.

3. Galler, Paul E., Lowenthal, Mark M., and Smith, Marcia S. "The Strategic Defense Initiative and United States Alliance Strategy." Report no. 85-48F. Congressional Research Service, Library of Congress, February 1985.

Addresses concerns of the European allies and the Japanese, on a case-by-case basis, with regard to the SDI and the increased likelihood of war, if it were actually deployed. Expresses fear of "making the world safe for conventional warfare."

4. Ottaway, David B. "Agreement Near With Bonn and London on SDI Research." *Washington Post,* September 17, 1985.

Describes breakthrough when Germany and United Kingdom announce plans to participate in SDI research. Assesses the impact on the Soviet Union as well as Congress.

5. Pierre, Andrew J. "An Irritant to Japan?" *New York Times,* July 11, 1985, p. 23.

States how the Japanese see SDI in terms of their friendship with President Reagan and the non-nuclear principles adopted after World War II. A caveat against pressuring the Japanese in a manner parallel to trade.